THE HUMAN DAWN

TimeFrame

4.6 billion BC

ARCHEOZOIC ERA
4.6 billion to 570 million BC

250 million BC

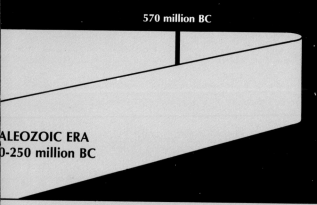

570 million BC

ALEOZOIC ERA
0-250 million BC

MESOZOIC ERA
250-65 million BC

CENOZOIC ERA
65 million BC to present

FIRST MEMBERS
OF THE GENUS
HOMO

65 million BC

2.3 million BC

TIME LIFE BOOKS

THE HUMAN DAWN

TimeFrame

BY THE EDITORS OF TIME-LIFE BOOKS

TIME-LIFE BOOKS, ALEXANDRIA, VIRGINIA

Time-Life Books Inc.
is a wholly owned subsidiary of
THE TIME INC. BOOK COMPANY

President and Chief Executive Officer:
Kelso F. Sutton
President, Time Inc. Books Direct:
Christopher T. Linen

TIME-LIFE BOOKS INC.

EDITOR: George Constable
Director of Design: Louis Klein
Director of Editorial Resources:
Phyllis K. Wise
Director of Photography and Research:
John Conrad Weiser

EUROPEAN EDITOR: Ellen Phillips
Design Director: Ed Skyner
Director of Editorial Resources:
Gillian Moore
Chief Sub-Editor: Ilse Gray
Assistant Design Director: Mary Staples

PRESIDENT: John M. Fahey, Jr.
Senior Vice Presidents: Robert M.
DeSena, Paul R. Stewart, Curtis G.
Viebranz, Joseph J. Ward
Vice Presidents: Stephen L. Bair,
Bonita L. Boezeman, Mary P. Donohoe,
Stephen L. Goldstein, Andrew P. Kaplan,
Trevor Lunn, Susan J. Maruyama, Robert
H. Smith
New Product Development: Trevor Lunn,
Donia Ann Steele
Supervisor of Quality Control: James King

PUBLISHER: Joseph J. Ward

Correspondents: Elisabeth Kraemer-Singh
(Bonn); Christina Lieberman (New York);
Maria Vincenza Aloisi (Paris); Ann
Natanson (Rome). Valuable assistance
was also provided by: Sasha Isachenko
(Moscow); Josephine du Brusle (Paris);
Traudl Lessing (Vienna).

TIME FRAME
(published in Britain as
TIME-LIFE HISTORY OF THE WORLD)

SERIES EDITOR: Charles Boyle

Editorial Staff for *The Human Dawn:*
Editor: Chris Middleton
Designer: Lynne Brown
Writers: Chris Farman, Christine Noble
Researchers: Marie-Louise Collard, Tim
Fraser, Louise Tucker
Sub-Editor: Luci Collings
Design Assistant: Rachel Gibson
Editorial Assistant: Molly Sutherland
Picture Department: Amanda Hindley
(administrator), Zoë Spencer (picture
coordinator)

Editorial Production
Chief: Samantha Hill
Traffic Coordinator: Emma Veys
Editorial Department: Theresa John,
Debra Lelliott

U.S. EDITION

Assistant Editor: Barbara Fairchild
Quarmby
Copy Coordinator: Ann Lee Bruen
Picture Coordinator: Barry Anthony

Editorial Operations
Production: Celia Beattie
Library: Louise D. Forstall

Computer Composition: Deborah G. Tait
(Manager), Monika D. Thayer, Janet
Barnes Syring, Lillian Daniels

Special Contributors: Stephen Downes,
Ellen Galford, Roy Hayward, Michael
Kerrigan, Alan Lothian (text); Judy
Aspinall, Barbara Moir Hicks, Rebecca
Hunter (research); David E. Manley
(index).

CONSULTANT

CHRIS SCARRE, Editor of the Cambridge
Archaeological Journal, Macdonald Institute for Archaeological Research, Cambridge, England

**Library of Congress Cataloging in
Publication Data**

The Human dawn / by the editors of Time-Life
Books.
    p.   cm. — (Time frame)
    Includes bibliographical references (p.   )
and index.
    ISBN 0-8094-6479-9—ISBN 0-8094-6480-2
    1. Human evolution.
    I. Time-Life Books. II. Series.
GN281.H8473 1990
573.2—dc20   90-11325
                              CIP

Time-Life Books Inc. offers a wide range of fine
recordings, including a *Rock 'n' Roll Era* series.
For subscription information, call 1-800-621-
7026 or write Time-Life Music, P.O. Box C-
32068, Richmond, Virginia 23261-2068.

CONTENTS

PROLOGUE TO HUMANKIND

Most of our knowledge about life before the advent of human beings is derived from fossils—traces of ancient life embedded in rock. The fossilization process began when animals and plants died and became buried in sediment such as mud or sand. The water and bacteria in the sediment then slowly dissolved the dead organisms, while at the same time impregnating the hard parts with mineral deposits. Eventually, a shape formed that was an exact replica of the original organism but was made of mineral rather than organic matter. Over millions of years, the sediment was drained of water and was compacted—along with the fossilized shape inside it—into solid rock.

Of the creatures that lived during the first four billion years of the earth's existence (chronology below), most left behind no trace; their remains were not made of sufficiently durable material. However, around 500 million BC, there emerged creatures with hard skeletal structures that did not decay immediately after death. These included vertebrates—animals with a backbone. Their fossil record begins with traces of bony fishes that had neither fins nor jaws; they probably wriggled through the water like tadpoles and fed on plankton. Later species developed teeth, hinged jaws, and fins, the most spectacular example being sharks, which appeared around 375 million BC.

From bony, finned fishes developed amphibians, the first vertebrates to walk on land. Like fish, these creatures laid their eggs in the water, but unlike fish, they were able to live on land as well. Pioneers of this lifestyle are thought to have been the *Crossopterygii*, fish that were possessed of muscular fins with which they dragged themselves across land and lungs with which they took in air directly.

Over time, amphibians developed legs and feet that operated more efficiently on land than did fins. Some smaller species began to spend less time in and around water, having been driven inland by larger rivals, and they developed characteristics reflecting this drier environment. They grew harder, scalier skins, and they laid eggs that had shell coatings to protect the embryo inside from desiccation. These were the first reptiles. The fossil on the right is that of *Seymouria baylorensis*, a thirty-inch-long creature that had a strong reptilian backbone fused to the pelvis but also possessed distinct amphibian characteristics—including labyrinthine teeth and, at either side of the back of the skull, a notch into which the eardrums fit.

Reptiles did not stay small. *Brachiosaurus*, the largest of all the dinosaurs, weighed almost ninety tons and measured more than eighty feet from head to tail. But around 65 million BC, the dinosaurs died out, for mysterious reasons—possibly the collision of a giant meteor with the earth, possibly because the reptiles, being cold-blooded and dependent on the sun for energy, were vulnerable to sudden drops in temperature. Their mastery over the earth passed to mammals, much smaller, warm-blooded creatures that suckled their young (the Latin word for "breast" is *mamma*.) These creatures also had evolved a variety of internal mechanisms for maintaining a steady blood temperature, which meant

ARCHEOZOIC ERA
(4.6 billion to 570 million BC)

PRECAMBRIAN LIFE
Formation of the earth (4.6 billion BC)
First life (algal limestone pillars) (3.3 billion BC)
First animals (sea worms, jellyfish) (680 million BC)

PALEOZOIC ERA
(570-250 million BC)

CAMBRIAN (570-510 million BC)
Invertebrates evolved (sponges, mollusks, shellfish)

ORDOVICIAN (510-435 million BC)
First vertebrates (500 million BC)
First fish (470 million BC)

SILURIAN (435-410 million BC)
First plants on land (412 million BC)

DEVONIAN (410-355 million BC)
First insects (380 million BC)
First amphibians (360 million BC)

CARBONIFEROUS (355-290 million BC)
First tropical forests (330 million BC)
First reptiles (300 million BC)

PERMIAN (290-250 million BC)

MESOZOIC ERA
(250-65 million BC)

TRIASSIC (250-205 million BC)
Conifers and ferns proliferated
First dinosaurs (215 million BC)

JURASSIC (205-140 million BC)
Continents split
First mammals (190 million BC)
First birds (140 million BC)

CRETACEOUS (140-65 million BC)
Many species died out, including dinosaurs (65 million BC)

that they relied not on the sun but on food for their energy.

From these early, mouse-size creatures evolved larger mammals, among them a group of tree-dwelling animals with long tails, grasping paws, and eyes at the front rather than at the sides of the head. The last feature greatly improved the judgment of distance, essential for an animal moving quickly from branch to branch. These were primates, from which, more than 20 million years ago, evolved the first apes.

The journey from bony fishes to apes spanned three distinct evolutionary periods—the Paleozoic, the Mesozoic, and the Cenozoic—the eras of old, middle, and new life, respectively *(chronology, below)*. The fossils of some of the creatures that lived in those eras are shown on the following pages. The evolution of their distinctive physical features shows the effects on those animals of the changing configurations of the continents that took place during this period—a phenomenon caused by continual movement of the earth's landmasses across the ocean floor.

CENOZOIC ERA
(65 million BC to present)

PALEOGENE (65-25 million BC)
Many mammals evolved (65-55 million BC)
First primates (60 million BC)
First horses (55 million BC)
First sea mammals (whales) (54 million BC)

NEOGENE (25-2 million BC)
First apes (23 million BC)
First *Australopithecus* (5 million BC)
First *Homo* (2.3 million BC)

QUATERNARY (2 million BC to present)
Most animal species in existence
First civilizations (8000 BC)

FROM FISH TO REPTILE

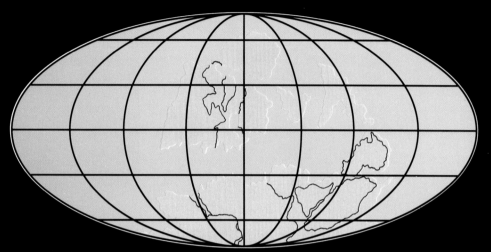

In the middle of the Paleozoic era, one giant continent—made up of present-day South America, Africa, India, Australia, and Antarctica—occupied the Southern Hemisphere *(map, left)*, while to the north lay the landmasses that would become North America, Europe, and Asia.

In the deep seas around the continents, vertebrate fish thrived. However, about 400 million BC, geological convulsions created new mountain ranges and lakes—habitats that attracted first plants, then insects and fish onto dry land. By the end of the era, some amphibians had evolved into exclusively land-based reptiles.

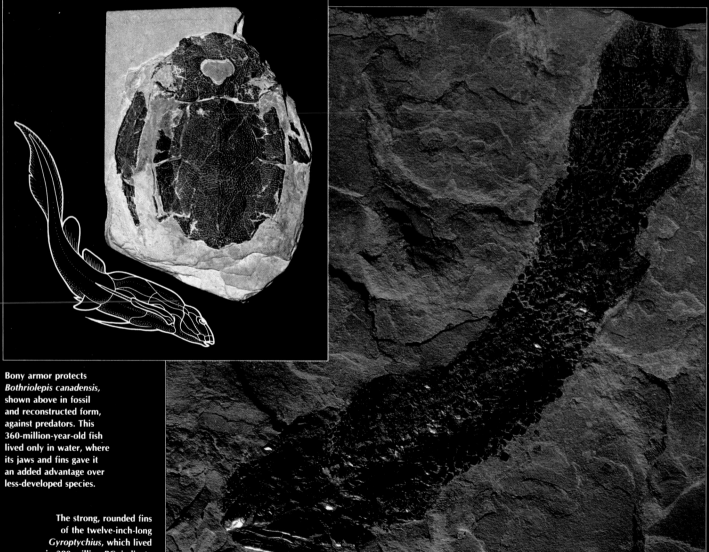

Bony armor protects *Bothriolepis canadensis,* shown above in fossil and reconstructed form, against predators. This 360-million-year-old fish lived only in water, where its jaws and fins gave it an added advantage over less-developed species.

The strong, rounded fins of the twelve-inch-long *Gyroptychius,* which lived in 380 million BC, indicate a fish that is evolving into an amphibian. Such fins slowly became limbs.

The fins have become fully formed legs in this fossilized imprint of a completely amphibian *Branchiosaurus,* which lived 300 million years ago. However, the creature retains its fishlike tail for use in the water.

Measuring little more than eighteen inches from head to tail, this *Labidosaurus hamatus* was one of the world's first reptiles, living entirely on land at the end of the Paleozoic era.

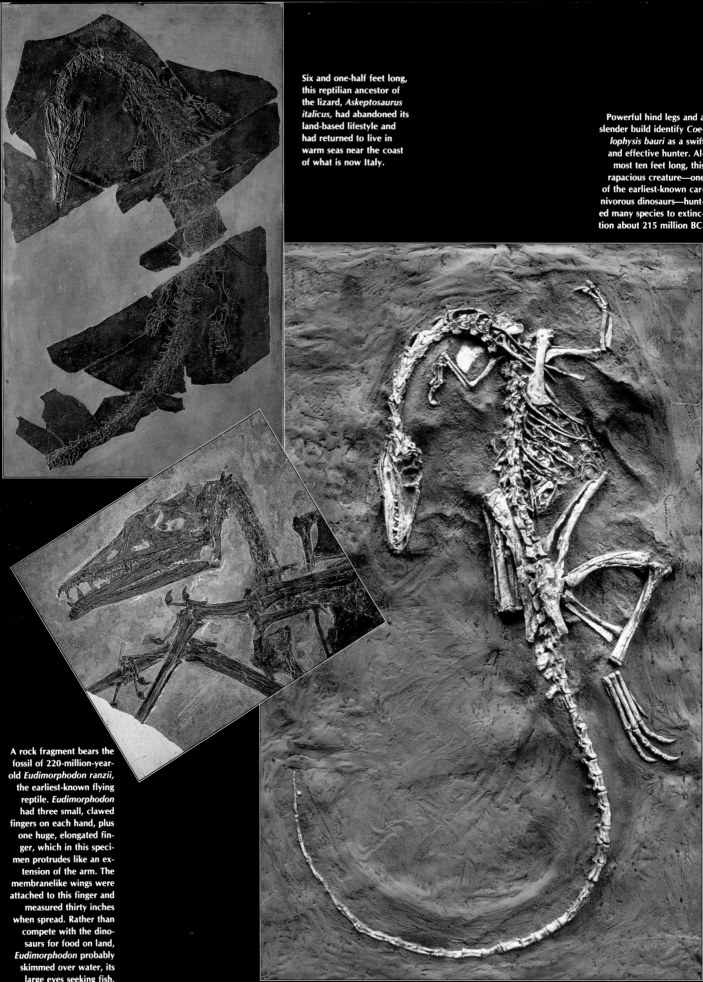

Six and one-half feet long, this reptilian ancestor of the lizard, *Askeptosaurus italicus*, had abandoned its land-based lifestyle and had returned to live in warm seas near the coast of what is now Italy.

Powerful hind legs and a slender build identify *Coelophysis bauri* as a swift and effective hunter. Almost ten feet long, this rapacious creature—one of the earliest-known carnivorous dinosaurs—hunted many species to extinction about 215 million BC.

A rock fragment bears the fossil of 220-million-year-old *Eudimorphodon ranzii*, the earliest-known flying reptile. *Eudimorphodon* had three small, clawed fingers on each hand, plus one huge, elongated finger, which in this specimen protrudes like an extension of the arm. The membranelike wings were attached to this finger and measured thirty inches when spread. Rather than compete with the dinosaurs for food on land, *Eudimorphodon* probably skimmed over water, its large eyes seeking fish.

THE EVOLUTION OF MAMMALS

By 170 million BC, the southern landmass had joined with its northern counterparts *(map, right)*, creating one vast continent and leaving a great gulf to the east.

Warm, rather humid conditions prevailed throughout this Pangaea, or "whole world." Lush tropical forests flourished, and in them thrived giant, leaf-browsing dinosaurs, who with their carnivorous counterparts came to dominate all other land vertebrates. Some smaller reptiles took to the air, others to the sea.

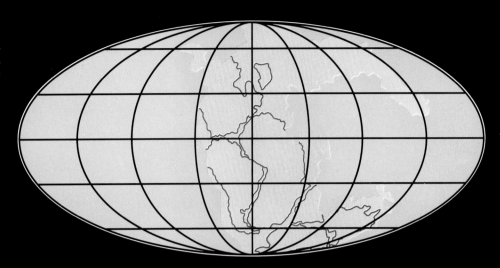

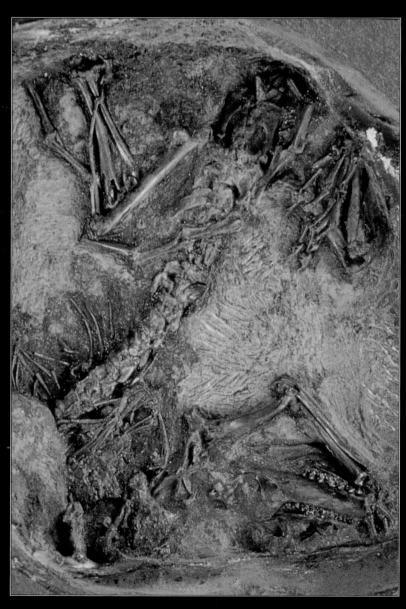

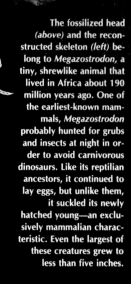

Based on fossilized remains *(left)*, a drawing of a pantothere *(right)*— the direct ancestor of almost all later mammals— shows an eight-inch-long creature well adapted for tree life, with a long tail for balance and powerful hind legs with which to launch itself from branch to branch. Like *Megazostrodon*, the pantothere avoided dinosaurs by nocturnal feeding.

The fossilized head *(above)* and the reconstructed skeleton *(left)* belong to *Megazostrodon*, a tiny, shrewlike animal that lived in Africa about 190 million years ago. One of the earliest-known mammals, *Megazostrodon* probably hunted for grubs and insects at night in order to avoid carnivorous dinosaurs. Like its reptilian ancestors, it continued to lay eggs, but unlike them, it suckled its newly hatched young—an exclusively mammalian characteristic. Even the largest of these creatures grew to less than five inches.

THE APPEARANCE OF APES

By 50 million BC, the world was beginning to look as it does today *(map, right)*. Australia was still joined to Antarctica, and India was adrift in midocean, but the Americas had separated from Africa and Europe. This, the Cenozoic era, saw the formation of the Alps, the Himalayas, the Rockies, and the Andes, as well as the evolution of flora and fauna similar to those we know today. A warm climate stimulated the growth of forests, in which mammals freely multiplied once the threat from dinosaurs had been removed.

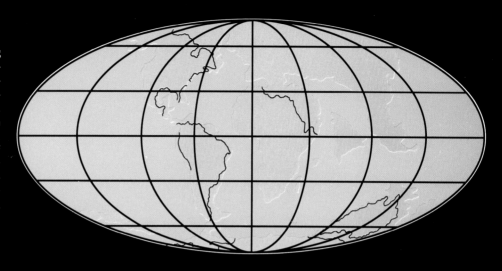

A restored skeleton of the 50-million-year-old, lemurlike *Smilodectes (left)* exhibits the characteristic features of a primate: grasping hands, enlarged brain area, and a snout that had grown shorter because an acute sense of smell was not essential for this tree-living animal, which fed only on fruit and insects. *Smilodectes'* eyes are positioned at the front of the skull, permitting the visual fields to overlap. These eyes transmitted a single picture to the brain, making it easy to judge distance; animals with eyes on the sides of the skull saw two images.

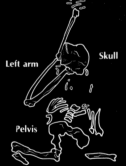

Right upper forearm and hand

Left arm

Skull

Pelvis

A fossil *(left)* found in a Tuscan coal pit bears the imprint of *Oreopithecus,* or "mountain ape," which lived toward the end of the Cenozoic era, about 14 million BC. Above, a drawing based on the fossil shows that although *Oreopithecus* had an ape's long arms, it had a pelvis broad enough to sustain a semiupright posture. In this fossil, humans are first foreshadowed.

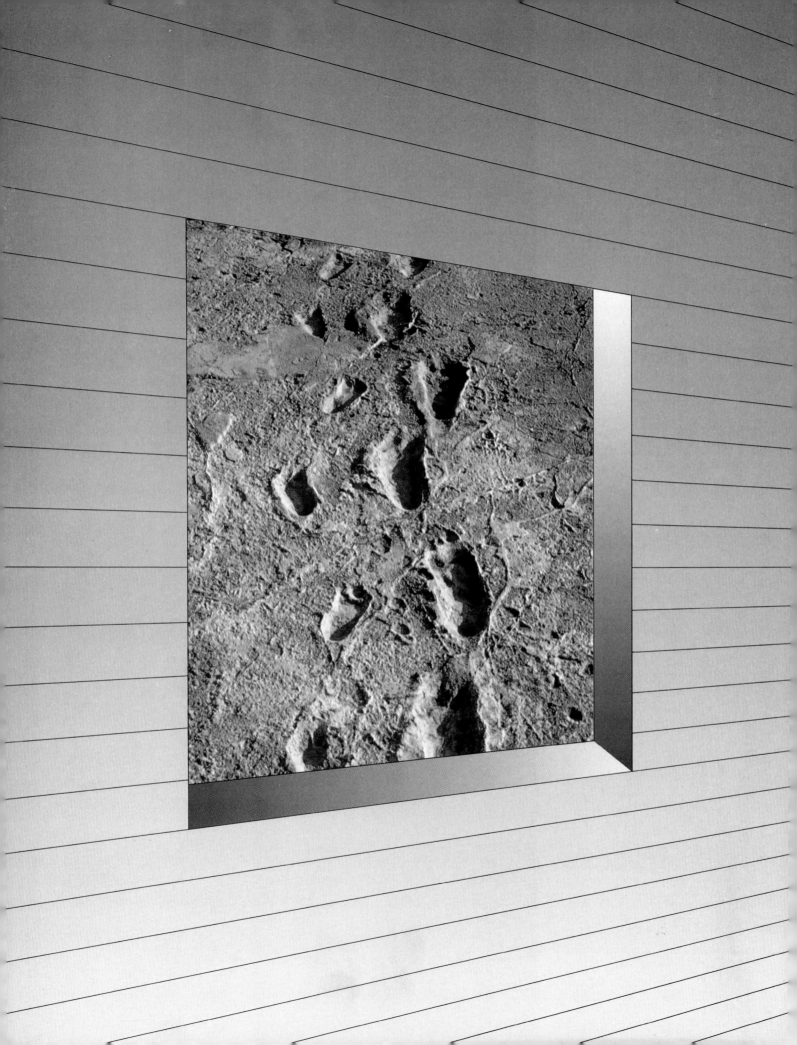

THE PATH OF EVOLUTION

1 On a warm afternoon in Africa, 3.5 million years ago, something in the long grass stirred. Antelopes at the lake's edge stopped drinking and looked up, their nostrils sifting the air for scent. They scanned the surrounding countryside, searching first the open savanna, its pale strands of grain shuddering in the gentle breeze, then the darker woodland behind, which stretched far up into the foothills before giving way to bald, parched mountain slopes. The only movement visible was high above, where a billowing white cloud traveled serenely across the sky, its aerial path mirrored on the ground by a large, slow-moving pool of shadow.

What was about to take place was never witnessed by human eyes, for the simple reason that 3.5 million years ago, human beings still did not exist. The only spectators at this re-created scene were the small clusters of birds and animals that had come down to the lake that afternoon to drink.

Whatever it was in the grass moved again. A group of gazelles nearby, sighting the disturbance, shifted nervously in the lake's muddy shallows. There was silence for a second, then, rising timorously above the line of light brown stalks, there appeared a tiny, dark brown head, in which a pair of eyes darted cautiously from side to side. The features that framed those eyes were those of an ape: flat-topped skull, low forehead, bony ridge over the eyes, nose flattened against the face, and the jaws spread wide across an almost horizontal chin. But there was something decidedly different about this creature, something strangely steady and measured about the way that it moved through the grass, showing no signs of the usual arm-dangling, lopsided gait that had fitted apes so well for life in the trees, but had always rendered them clumsy and vulnerable when down on the ground.

Slowly yet purposefully, the animal advanced. As it reached the edge of the long grass, it paused once more to scan the surrounding countryside for possible predators. Then, with one bold step, it emerged from the undergrowth, not crawling on all fours but walking straight and upright on two feet, its arms not trailing along the ground but gently swinging at its side. Without the cover of the savanna, its diminutive stature became instantly apparent; although a fully developed female of about twenty years, the creature stood little more than four feet tall from the top of her hairy head to the tips of her long, slender toes.

To the gazelles and antelopes at the lake's edge, this lone, diminutive figure warranted no more than a passing glance; right away they assessed her as presenting little threat to them and returned to their drinking.

But this small creature venturing out for a drink some 3.5 million years ago, in the Afar region of eastern Ethiopia, represents a pivotal point in the whole epic saga of humankind's evolution from apes to human beings. For she belonged to a group of mammals known as *Australopithecus afarensis,* literally the southern apes from Afar,

Footprints offer incontrovertible evidence that human ancestors strode across the Laetoli region of Tanzania under the East African sun more than 3.5 million years ago. The largest prints, thought to be those of an adult male, were overlaid by steps of a smaller hominid—possibly a female following in his wake—while a younger companion, perhaps a child, walked alongside. As the savanna season turned from dry to wet, the three walkers left their tracks in a layer of volcanic ash, which rain had turned to mud; after they passed, the sun baked their footprints hard. Later, ash covered and preserved these marks of the genus *Australopithecus,* the first hominids to walk fully erect, leaving their hands free.

who were the first creatures known to have been fully bipedal—that is, to walk on two feet as the rule rather than as the exception. Furthermore, although these australopithecines usually lived and slept out in the open air, and although they shared a great many of their physical characteristics with gorillas, baboons, and chimpanzees, they stood on an evolutionary path that was eventually to lead to modern-day humans. Thus, despite the fact that the Ethiopian water seeker may have been naked and hairy and may even have gotten down on all fours to drink from the lake, she can claim direct ancestry with her modern-day counterpart.

Indeed, had they but known it, the animals gathered at the lakeside that day were looking at a creature whose descendants would inherit the earth. Although the female australopithecine would be dead within ten years—few of them lived past thirty—the generations that followed her would develop brains that grew ever larger and hands that grew ever more dexterous. It was this mutually enriching interaction between mental and manual skill that over the next three million years would fuel her successors' genetic journey from apes to anatomically modern human beings, and would in the process enable them to create an armory of weapons, tools, and tactics that would ensure their eventual domination of every other species on the earth.

The planet on which the australopithecines were struggling to gain a foothold had already been in existence for many millions of years. According to the current estimates of astrophysicists, it was approximately 15 billion years ago that a cosmic fireball of inconceivable magnitude suddenly exploded. Multitudes of galaxies were formed out of the fragments of this explosion, and in one of these, clouds of gas condensed and formed a medium-size star, our sun. Other gas clouds, circling around this star, condensed too, forming a number of planets about 4.6 billion years ago. One of these planets was our earth.

The enormous, mind-boggling timespan involved is, for most people, an obstacle to understanding the events of the remote past. It was so much easier when scholars could simply accept the calculations of James Ussher, an Irish archbishop who, in 1650, added up the number of generations recorded in the Bible and pronounced that the world had been created in the year 4004 BC. This estimate was later refined by Bishop John Lightfoot, vice-chancellor of Cambridge University, who further narrowed the hour of humanity's creation down to 9:00 a.m. on October 23, 4004 BC. Nevertheless, it is possible to put the story of evolution into some kind of manageable perspective by adopting a rough time frame that condenses the 4.6 billion years of the earth's past into one calendar year.

Working on this basis, the earth's atmosphere would have been unbreathable for the first seven months. It was only after ten months had passed that anything larger than microscopic organisms started appearing, when the interaction of sunlight and bacteria began releasing oxygen into the atmosphere—the process known as photosynthesis. Fish followed two weeks later, and seven days after that, plants started growing on land. The next nine days saw much activity, with the arrival of insects, amphibians, forests, and reptiles, and it was just fifteen days ago that the earth first felt the ponderous tread of dinosaurs.

Mammals followed the next day and, for some as yet unknown reason, managed to survive when the dinosaurs, possibly due to the earth being devastated by a meteorite, suddenly vanished a mere five days ago. If we take the present time as being noon on a Monday, then the first apes arrived on the previous Friday afternoon, the australopithecines at 6:00 this morning, and anatomically modern humans, known to scientists as *Homo sapiens sapiens,* emerged only eleven minutes ago. More recently, it has been one minute since the birth of agriculture, 20 seconds since the building of the first city, and only 1.5 seconds since the signing of the American Declaration of Independence.

The creation of the world and its myriad inhabitants constitutes a complex evolutionary story, but within the chemical and biological compositions of the creatures that participated in it, there occurred developments of a less visible but no less significant nature. The first of these major events was the creation of life itself, which took place around 3.3 billion years ago. On this subject, nothing can be said with any degree of certainty, although much can be surmised by tracing genetic changes backward from life forms that exist today. For example, it is known that all living creatures are composed of cells—small, separate units, each of which has the capacity to reproduce itself, thanks to the presence within its confines of a substance commonly known as DNA, or deoxyribonucleic acid. Each cell contains, in the form of a double helix, two intertwined strands of DNA molecules, and it is the quantity and configuration of those molecules that provide the blueprint for the individual cell's genetic characteristics.

In order for a cell to create a replica of itself, it must assemble a new, matching pair of strands. If the assembly is accurate, this new cell will be an exact copy of the original. In this way, cells have the capacity to continue making copies of themselves ad infinitum. However, nature has never been so benevolent as to permit indiscriminate and continuous reproduction. Living things might have the right to be born, but nature has always been more sparing when it comes to the right to survive. In prehistoric times, as now, survival was something that a species could achieve only through a gradual process of adapting to its circumstances. And the only long-term

APES AND HUMANS

For countless millennia, human and ape traveled a common evolutionary road, until, around five million years ago, human precursors stood upright. This development demanded the growth of new bone and muscle structures. At a later stage, and even more important, the brain developed as a control center for new patterns of movement and behavior. As illustrated on page 23, the progression from early hominid to modern human was not a direct line: The influence of a changing environment on competing, coexisting species led to evolution for some, extinction for others. Today, *Homo sapiens sapiens* is the only surviving hominid line, to which the closest relatives, by virtue of their genetic similarities, are the chimpanzee and the gorilla.

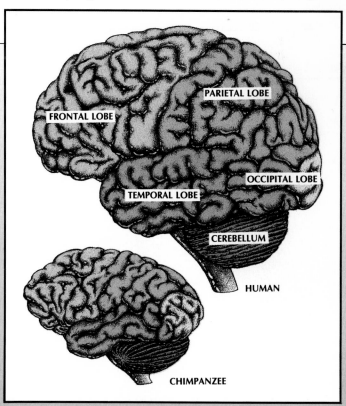

The superiority of the human brain over that of the ape lies not only in its size—three times that of our nearest ape relative, the chimpanzee—but also in its shape. The areas responsible for muscular coordination (cerebellum), higher mental capacities such as memory, intelligence, and personality (frontal lobe), interpretation of sensory information (parietal lobe), and hearing and speech (temporal lobe) are all more highly developed in humans. Only the occipital lobe (responsible for vision) is more developed in the chimpanzee.

FRONTAL LOBE

PARIETAL LOBE

OCCIPITAL LOBE

TEMPORAL LOBE

CEREBELLUM

HUMAN

CHIMPANZEE

means of attaining such an adaptation—to circumstances that themselves were constantly shifting—was through genetic change.

Surprisingly, this change would be initiated in many cases by an accident. Then, as now, it sometimes happened that the DNA copying process would not be entirely accurate, with the result that the new organism inherited a characteristic or characteristics dramatically different from the original organism. Mostly, the changes would be for the worse, and the new mutant organism would perish. Sometimes the change made no appreciable difference. But, occasionally, a mutant creature would be capable either of doing some things better than its predecessors or, in a few instances, of doing things its predecessors just could not do.

For example, a creature would be born possessing a new characteristic—such as a sharper tusk—which, although differentiating it from the rest of the species, would nevertheless prove to be such an asset that its bearer became more genetically successful than its counterparts: That is, it lived longer and bred more offspring than they did. As time wore on, and the success story repeated itself from generation to generation, the sharper-tusked creature became the norm, until it, too, was replaced by another mutant strain, perhaps one with shorter, more agile legs.

This capacity for genetic improvement was perfectly embodied by the Ethiopian australopithecine down by the lake's edge. For although her lineage stretched directly forward to large-brained, manually dexterous modern humans, it also stretched back to small-brained, ham-handed creatures, who stood with all four limbs firmly in the ape camp. Thirty million years ago, in the tropical rain forests of Africa, monkeys lived much as mon-

Although the hands of apes and humans display obvious structural similarities, the human thumb is not only longer but also has a far greater range of movement than that of the ape. By contrast, the flat of the ape's hand is larger than the human's, giving it a more powerful grip but less digital dexterity.

keys do today. They spent their days high up in the leafy canopy, scampering along the branches and using their long tails to steady themselves either when they stood still on all fours or when they leaped from tree to tree in search of fruit. However, at about the same time, a different breed of monkeys, somewhat larger in size, took advantage of their longer arm span to initiate a new form of food gathering. Their technique was to dangle. By simply attaching one hand to a branch and then lowering themselves in the right direction, they could get hold of fruit that had previously been out of reach. Gradually, the inventors of this aerial harvesting method progressed from passive dangling to active swinging, in the process developing wrists that were more mobile and arms that were longer still. And since they now provided their own means of balance, their tails not only became redundant but actually disappeared. Tailless, they could no longer be classified as outsize, unconventional monkeys. They belonged to a new category altogether: apes.

Slowly, their new lifestyle brought about further changes. Perhaps because of a different diet that required more chewing and grinding, apes developed sturdier molars, with five cusps, or points, compared with the monkeys' four. Gradually, too, their posture became more upright than that of the monkeys, and their ability to sit rather than crouch freed their hands for the more intricate tasks of plucking, poking, holding, stroking, and examining. This gave them a clear advantage over monkeys, who could grip a nut or a small piece of fruit in their hands, but if they had to make a quick getaway, needed all four paws for running and would have to drop their food. Apes, on the contrary, could make off with their prize still clasped in one hand. As a bonus, it also turned out that the more the apes used their hands, the more intelligent they became; increased manual dexterity stimulated the growth of their brains, which, in turn, equipped them with the necessary intelligence to perform yet more complicated tasks.

And there they might have stayed, enjoying a comfortable superiority to the monkeys and swinging contentedly through seemingly limitless forest. But it was not to

Bone for bone, the human skeleton corresponds with that of the ape, but the demands of upright walking produced significant changes in shape. The spine developed a curve to support the vertical posture; the pelvis became shorter and basin-shape to accommodate the stance and gait; and the legs lengthened. Hominid behavior and diet reduced the need for prominent canines—used by apes in displays of hostility—and caused the gaps between the teeth to disappear.

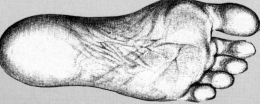

Long toes and a divergent big toe adapt the gorilla's foot to gripping, while the shortened toes of humans promote upright stability. The gorilla lacks the human arch, which absorbs into its vaulted structure the impact of each step on the rest of the body.

be. In about 15 million BC, the world's climate began to change, slowly becoming more seasonal and more arid. In the great lush rain forests, where food had always been available, sparser deciduous trees came to dominate. In many places, the forest thinned away to savanna, which was nothing more than open grassland with scattered trees. Suddenly, the animals that hitherto had dwelt in the forest had to adjust to a new set of circumstances. Those who lived on the ground needed only to alter their diet; antelopes and horses, for example, switched from leaves to grass. However, those creatures whose homes had been the trees were forced to make more drastic changes if they were to survive.

The first ape that adapted successfully to life away from the forests was *Ramapithecus,* the Rama-ape, named after the Hindu god Rama. The fossilized remains of *Ramapithecus* are found as far apart as Spain and China, but they are at their most plentiful in the foothills of the Himalayas in Pakistan, where 13 million years of erosion from the mountains have left layers of sediment nearly two miles deep. Within these layers are preserved the jaws and skulls of several specimens of *Ramapithecus* that are between 10 and 12 million years old.

The Rama-ape was of modest dimensions—barely four feet tall—but what differentiated it most clearly from other apes was its teeth. The canines, which in earlier primates had been dagger-like, had in this species shrunk until they were not much larger than the incisors, and the molar crowns had developed thick, hardwearing layers of enamel. With this new dental arrangement, *Ramapithecus* could move its lower jaw from side to side, unimpeded by protruding canines, and grind its food between molars now strengthened for the task. These were not the jaws of a creature that fed on soft, mushy tropical fruits plucked from the treetops, but those of an animal that ate hard objects such as grass seeds and grains. At least one ape, it seems, had come down from the trees for good.

Unfortunately, there is a vast gap in the fossil record that covers the period during which the ancestors of human beings diverged from the African apes. For five million

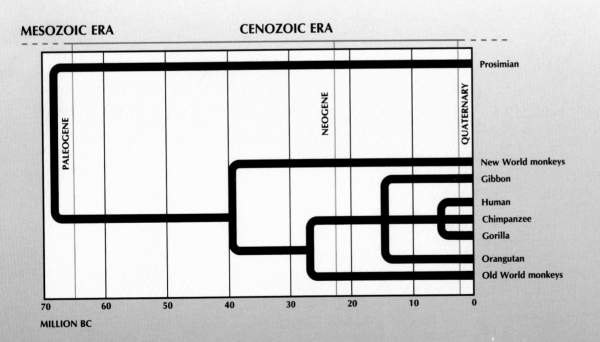

MESOZOIC ERA   CENOZOIC ERA

PALEOGENE   NEOGENE   QUATERNARY

Prosimian

New World monkeys

Gibbon

Human

Chimpanzee

Gorilla

Orangutan

Old World monkeys

70   60   50   40   30   20   10   0

MILLION BC

A DNA chart *(left)* comparing chromosome make-up shows the close relationship of humans to the chimpanzee and the gorilla, and the comparatively recent date at which their evolutionary paths diverged. DNA—deoxyribonucleic acid—is the main constituent of chromosomes, which are responsible for passing on hereditary characteristics in animals and plants. Human DNA differs by less than two percent from that of chimpanzees but by nearly eight percent from that of baboons. The evolutionary development chart *(right)* illustrates the gradual and indirect process that led to modern *Homo sapiens sapiens.* Areas shaded gray indicate dating still under debate.

years after the last *Ramapithecus,* between 10 million and 5 million BC, almost nothing has survived of the apes. Only an occasional five-cusped tooth has surfaced to show their continued presence in Africa and Asia. When the record is again clearly legible, there is evidence of strange new creatures, no more than four feet tall, with brains barely larger than those of chimpanzees. Their canines were small, and their molars were thick, but they possessed one characteristic that was more astonishing than all the rest put together. They walked upright.

These were the australopithecines, or southern apes, whose remains have been found at sites in both South and East Africa. The earliest australopithecines lived about five million BC, and the last disappeared from the earth in approximately one million BC. The fossils of these first bipeds showed a great number of anatomical changes from the characteristics of their ancestors. The hole in the base of the skull through which the spinal cord passed—known as the foramen magnum—faced almost directly downward, whereas in apes, it faced outward from the back of the skull. The spine, too, no longer formed a hunched arch, as in other primates of that time; instead, it was an *S*-shape pillar, erect in the middle, and curving slightly at the neck and in the small of the back. The pelvis had become shorter, broader, and more bowl-shaped, so as to support the weight of the torso above. The hip joints were modified so that the legs could align with the backbone, and the knees had grown closer together in order to bear the total body weight more efficiently. The ankles had become stronger and more inflexible than an ape's; the soles of the feet had grown vaulted rather than flat; and the big toe had become aligned with the other toes, thereby losing its ability to grasp, but acquiring the capacity to take the full weight of the body during each stride, as it still does today. Examination of footprints made by australopithecines at Laetoli in Tanzania shows that their gait exactly matched that of modern-day peoples who have never worn shoes.

Australopithecines varied quite markedly in their physical dimensions. Fully grown adults of the smallest species could measure as little as four feet in height and weigh

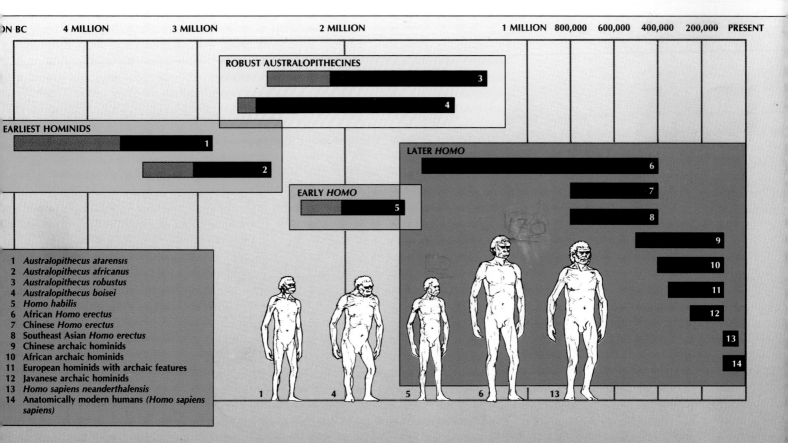

ROBUST AUSTRALOPITHECINES

3

4

EARLIEST HOMINIDS

1

2

LATER *HOMO*

6

7

8

9

10

11

12

13

14

EARLY *HOMO*

5

1 *Australopithecus atarensis*
2 *Australopithecus africanus*
3 *Australopithecus robustus*
4 *Australopithecus boisei*
5 *Homo habilis*
6 African *Homo erectus*
7 Chinese *Homo erectus*
8 Southeast Asian *Homo erectus*
9 Chinese archaic hominids
10 African archaic hominids
11 European hominids with archaic features
12 Javanese archaic hominids
13 *Homo sapiens neanderthalensis*
14 Anatomically modern humans (*Homo sapiens sapiens*)

not more than sixty-five pounds, whereas the biggest members of the largest species stood up to five feet tall and weighed 155 pounds. However, they did possess some notable features in common: All of them had large, well-developed molars and brains that were much smaller than those of modern-day human beings. The average capacity of an adult australopithecine's cranium—the part of the skull housing the brain—was about 450 cubic centimeters (27.5 cubic inches), 65 cc more than that of a chimpanzee, but nearly 1,000 cc less than that of an average person today.

Few clues remain as to the australopithecines' lifestyle. No tools, for example, have ever been found that might indicate the range of routine tasks they performed. Raymond Dart, the Australian-born anatomist who made the first australopithecine discoveries in South Africa in 1924, originally supposed that these first hominids lived

Virtually unchanged from its ancestors of 400 million years ago, a modern-day coelacanth *(right)* drifts through the deepest reaches of the Indian Ocean. Whereas environmental change forced the evolution of new characteristics among most other species, the coelacanth seems to have possessed the inbred capacity to adapt to that change without developing dramatically different features. Other life forms that have undergone similarly few changes over many millions of years include the bat, the turtle, the opossum, and the cycad, a palmlike plant on whose cones dinosaurs once fed.

by hunting, because the first bones were found in caves among the remains of antelopes and other mid-size prey. Later, however, scientists realized that these australopithecines were the prey, too. Instead of dying alongside the bones of the beasts they had killed, they themselves had fallen victim to some large animal—perhaps a leopard—which had devoured them and then discarded their bones.

Judging by their large molars, ideal for grinding coarse plant and vegetable foods, the australopithecines seem to have lived as herbivores, perhaps going up into the trees at night for safety, but spending the day down on the grasslands and lakesides of Africa. They shared their environment both with the ancestors of animals that still live there today and with some—such as the giant grass-eating baboon, almost six feet tall—that have since become extinct.

Most anthropologists agree that there were four main species of australopithecines. A representative of the earliest species—*Australopithecus afarensis*—has already been mentioned, in the re-created scene by the Ethiopian lakeside, where she and her companions lived approximately 3.5 million years ago. The American anthropologist Donald Johanson discovered her fossilized remains in 1974, when a flash flood washed away the walls of a gully to reveal a piece of her armbone. The next species for which we have firm evidence is *Australopithecus africanus,* the first skeletal remains of which were discovered under even more dramatic circumstances with the aid of gelatin dynamite.

One day in 1924, at the Taung limestone quarry near Johannesburg, the day's blasting was continuing as usual when, after one explosion, workers inspecting the site came across two most unusual pieces of rock. Embedded deep in one of these pieces was half a skull which, judging by its size, was that of a small child, with its milk teeth still in place. Set into the other piece was the fossilized cast of this same child's brain, with all its convolutions and furrows still clearly visible. Sensing immediately the significance of their discovery, the workers took the rocks to the quarry manager, who then sent them on to Raymond Dart. After using—among other implements—his wife's knitting needles to pick away the lumps of surrounding sediment, he carefully examined the small skull and declared that it belonged to a member of a hitherto-unknown species. He named it *Australopithecus africanus*, the southern ape of Africa.

Later finds of similar fossilized remains—also in South Africa—gave a clearer

picture of what *Australopithecus africanus* looked like. *Africanus*, somewhat taller and stronger than *afarensis*, inhabited the earth between three and two million BC. It stood upright and, although not very tall—four foot three at the most—nevertheless possessed a sturdy, muscular physique and weighed between 100 and 110 pounds. Compared with what is known about the lifestyle of *afarensis*, it seems to have enjoyed a more specialized plant diet. Its molars had become much larger, and its jaw had increased considerably in strength in order to facilitate more efficient grinding. However, its canines and incisors—the teeth that are at their most developed stage in carnivores—had grown smaller. This change indicates that the types of food eaten by *africanus* needed little preliminary tearing or slicing, and probably consisted only of a variety of fruits or seeds covered by hard pods or husks that had to be ground down before giving up their contents.

The trend toward plant eating continued in the two later australopithecine species, *Australopithecus robustus* and *Australopithecus boisei*, both of which appear to have lived between two million and one million BC, before becoming extinct. Remains of the former were unearthed in South Africa in 1938, and it certainly deserved its name: It stood no more than five feet tall, but a big male may have weighed as much as 155 pounds. Both its skull and jaws were very strongly built, with massive bone ridges that served to anchor the powerful chewing muscles that its diet required. These developments went even further in *Australopithecus boisei*, whose remains were discovered in 1959 at Olduvai Gorge, in modern-day Tanzania, by the archaeologists Mary and Louis Leakey. This species was named after the American-born London busi-

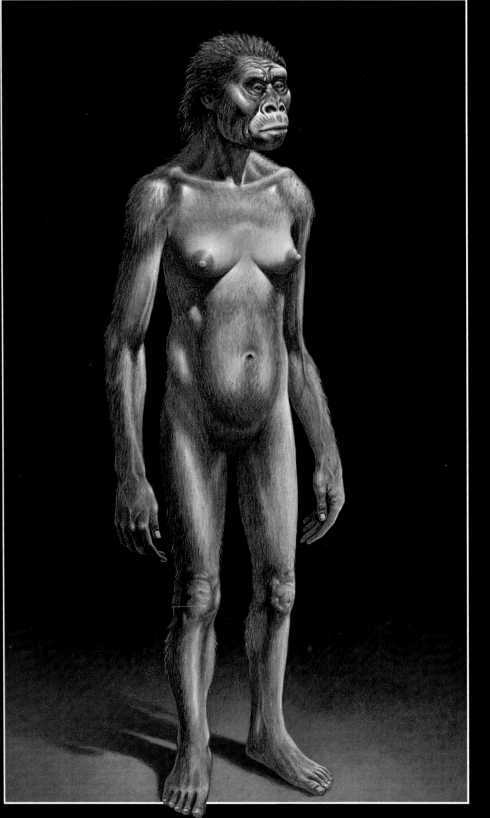

In this artist's reconstruction, "Lucy," a female hominid of the species *Australopithecus afarensis,* gazes out from the distant past of around 3.5 million years ago. Her skeleton was found in 1974 by anthropologist Don Johanson at Hadar in Ethiopia, and she was named after a song that was played over and over as Johanson and his colleagues celebrated the discovery—the Beatles' "Lucy in the Sky with Diamonds." Lucy was smaller than a woman of today *(above),* standing about four feet tall and weighing about seventy pounds. Although she was not human, Lucy had many humanlike features, the most notable being her erect stance and her hands.

nessman Charles Boise, who had funded the work undertaken by Mary and her husband. So large had this specimen's jaws become that their owner was popularly dubbed Nutcracker Man.

No one questions the fact that what unites the australopithecines with modern-day human beings is their ability to walk upright. But what no one has yet agreed on is why they should have chosen to do it.

There are obvious disadvantages. In biological and biomechanical terms, walking on the hind legs with the body held vertical is an extraordinarily complicated and difficult balancing act. Also—if judged on speed alone—it is a blatantly inefficient way of moving, since it makes use of only two limbs, whereas quadrupeds such as apes and monkeys—with four limbs at their disposal—can easily outdistance a human being. In most mammals, each limb makes a contribution to locomotive power, but in human beings, the arms are a dead weight. The only bipeds that can move quickly are those, such as ostriches and kangaroos, whose front limbs have become so small and weak that they are no impediment to running.

However, one major advantage that bipeds do have over quadrupeds is that they possess far higher levels of endurance. Whereas many human beings are capable of running a twenty-six-mile marathon, a zebra will fall down exhausted after being chased for less than 900 yards.

So what could have caused australopithecines to favor two-footed locomotion as a means of traveling on the ground? Other apes and monkeys have come down from the trees and yet have remained quadrupeds, standing upright only briefly in order either to get a better view or to reach a high fruit. Why not the southern apes?

Several factors must have combined to make walking upright the most efficient solution to the problems that they faced. One element may have been the australopithecines' need to present less of their bodies to the fierce tropical sun; an upright creature receives only two-thirds as much sunlight as does one on all fours, and even less when the sun is directly overhead.

Another contributory factor may have been an increased use of the hands, which could have come about for two reasons. First, the earliest primates communicated with one another through a primitive system of calls and, therefore, needed to devise a method of carrying food and objects other than in their mouths. The second possibility is that they needed to keep their hands free for performing manual tasks and also for transporting food back to their companions. Whereas the apes of today live in rather loose family groups, in which each individual ape feeds itself, the australopithecines might have evolved a system whereby they brought back the food that they foraged in order to share it with other members of their family. If so, then possessing a good carrying ability would have been vital for survival, and the most successful australopithecines would have been those who were able to master the combination of efficient conveyance (using their two upper limbs) and swift running (using their two lower limbs).

That they achieved this capability in significant numbers is demonstrated by the existence of humans today. Most experts tend to agree that modern-day human beings are descended from australopithecine stock, although they largely disagree over the nature of that descent, as, indeed, they disagree over the relationship between the various species of australopithecines themselves.

How, when, and where the changeover occurred from australopithecines to human beings—members of the genus *Homo*—is still far from clear. What we do know, however, is that *Australopithecus* and *Homo* existed alongside each other in Africa for a substantial amount of time, perhaps for as long as 800,000 years. Far from being doomed and brutish creatures, as they are often portrayed, the australopithecines at first seem to have achieved greater reproductive success than did their hominid counterparts: In excavations carried out at various African sites, archaeologists have unearthed almost twice as many fossilized remains of australopithecines as they have those of the genus *Homo*.

Contrary, then, to popular legend, the ape-men did not suddenly die out with the arrival of human beings, nor did human beings deliberately wipe out their more primitive forebears. Those australopithecines that did not follow the evolutionary path toward humankind became extinct solely because they failed to adapt to the increasingly competitive situation on the African savanna, where they found themselves overtaken, first by more efficient herbivores—such as the gazelle, the oryx, and the horse—and then by a growing number of species, including early *Homo,* that had incorporated meat into their diet. Squeezed between these more efficient herbivores on the one hand, and the more skilled omnivores on the other hand, the last of the australopithecines lost their battle for survival and finally disappeared from the face of the earth about one million BC.

But, while the australopithecines were on the evolutionary way down, there emerged on the earth, around 2.3 million BC, new creatures that were very much on the way up. These were humans: not apes or even ape-men, but the world's first full-fledged members of the genus *Homo*—meat eaters and grass eaters combined, who not only had brains that were more than 50 percent larger than those of the australopithecines but also possessed an ability to make their mark on the environment and to shape their daily existence in a way that no creature before had been able to. They could make stone tools.

Known as *Homo habilis,* Man the Skillful, their remains were first discovered, like those of *Australopithecus boisei,* at Olduvai, in eastern Africa. Judged by height and weight, they were not particularly remarkable—no more than four feet tall and 110 pounds in weight. But they had 800-cc brains and were capable of doing what no australopithecine is ever thought to have done: They split stones in order to make sharp-edged cutting tools. They probably did the same with wood as well, but unlike stone implements, these would have long since rotted, leaving no traces; however, it can be surmised that creatures with the intelligence to exploit rocks for their own ends must have done the same with trees.

From the cut marks on the bones of large mammals that have been found at *Homo habilis* sites, archaeologists can tell that one of the primary uses for these tools was to butcher animal carcasses. However, this does not mean that those early humans could acquire unlimited amounts of meat. Analysis of these bones shows that the parts to which *Homo habilis* had access were usually the least fleshy parts of the carcass. Often, too, they seem only to have eaten the leftovers; many of the tool marks overlap the much bigger incisions made by large-toothed carnivores, who presumably had already helped themselves to the choicest cuts of meat.

It seems certain, then, that meat was added to the *Homo habilis* diet, perhaps as a result of a dwindling supply of vegetation, caused either by seasonal shortages or by competition from other animals. It also seems clear that *Homo habilis* obtained meat not by confronting large beasts in close combat, but by a policy of cautious and opportunistic scavenging. This involved either picking up the scraps that larger creatures had left behind or using sharp tools to open up the corpses of animals whose hides were too thick for other carnivores to pierce. This latter method at least gave the hominids first bite at the carcass before being chased off by large predators, such as lions and panthers, or by any lesser animals, such as hyenas, that could gather in sufficient numbers.

Meat in the diet brought social as well as nutritional benefits. An animal carcass represented a substantial amount of food, usually more than one individual could manage, and it is likely that a pattern of behavior emerged in which *Homo habilis* shared feasts with their companions—if only because they could then expect favors in return. Although it hardly amounted to an organized system of food distribution, this partaking of communal meals must have provided these hominids with an opportunity for social activity and for the forming of a rudimentary sense of fellow feeling toward members of their own kind.

The pursuit of meat also expanded geographical frontiers. Whereas most plants could grow only in a place that offered the right conditions—light or dark, hot or cold, moist or dry—animals were, on the whole, less sensitive to the distinctions of shade and temperature and, consequently, were inclined to roam farther afield, drawing their human predators after them.

This increased level of exertion, beneath the tropical sun, may have been a contributory factor in bringing about human beings' relative lack of body hair. It seems quite possible that hominids chose to do their hunting and scavenging during the hottest times of day, when other, larger predators were resting; repeated over many thousands of years, this practice could have led to a gradual shedding of body hair and an increased dependence on sweating as a means of keeping body temperature down in the midday sun. Human beings perspire more profusely than any other mammal, even though their sweat glands are proportionally the same size as those of the others. And whereas other carnivores use panting as a method of cooling themselves, this is effective only after short bursts of exertion; over the length of time that scavenging hominids might need to chase a large animal until it collapsed, panting would have led to dangerous hyperventilation.

Sweating provided hominids with the most suitable evolutionary solution to high-temperature living. However, although it brought benefits, it also imposed a number of restrictions. The steady loss of body fluid meant that hunters had to confine their operations to areas where they knew that drinking water would be readily available. Humans cannot tolerate a water loss of more than 10 percent of their total body weight; and inasmuch as camels, for example, can stock up with large amounts of liquid, swallowing twenty-five gallons in ten minutes, human beings can manage only one quart in the same time frame. Proximity to water, therefore, became essential to those engaged in the active pursuit of meat.

Finding ways to meet these new challenges both stretched and stimulated the hominids' intelligence. Whereas for as long as three million years, the brain of an average australopithecine had remained the same size, proportionally, as that of a chimpanzee, the human brain now embarked upon a slow and continuous process of enlargement that would end only when it was more than three times bigger than that of the largest australopithecine.

Several complex and interrelating factors brought about this growth. Scavenging and hunting made more demands on the intelligence than eating grass, especially in

a species that did not possess the physical assets of a killing machine, such as the leopard or the lion. And access to a food like meat, which was rich in protein, provided human beings with the additional energy source they needed to fuel the development of a bigger brain.

The making of tools required a level of dexterity that also served to promote an increase in brain size. No one precisely understands the correlation, but it has been observed that animals exhibiting a degree of manual skill, such as the chimpanzee, are usually more intelligent than those that do not display such an ability. It can be assumed, therefore, that the very act of chipping stone to make tools must have literally broadened *Homo habilis'* mind, providing the intellectual capacity to perform yet more complex manual tasks, which in turn would have had a further nourishing effect on the brain.

Toolmaking also required a mental input that went well beyond that of merely knowing how to make the implement. Visualizing the existence of a sharpened edge within a rounded stone was an act that required a new level of imagination and insight. In addition, while *Australopithecus* merely needed to know where to go to find food, *Homo habilis* also had to know where to go to find the raw materials for toolmaking. This kind of activity necessitated foresight, memory, and planning in a way that plucking berries off a bush did not.

As the size of the human brain increased, so the head grew in order to accommodate it. Here, though, a problem presented itself. The new, upright walking position limited the width of the human pelvis; if it got any wider, the legs would become splayed and inefficient, thereby setting human beings back on the road to quadrupedism. However, this in turn limited the width of the female birth canal, thereby restricting to about 350 cc the maximum cranial capacity of a newborn baby, who had to pass through that canal during birth. Since most primates' brains only doubled in size between infancy and adulthood, this meant that the brain size of a full-grown adult would never grow past 700 to 800 cc. At this point, then, it seems that the evolutionary process should have ground to a halt. What had previously been human beings' inexorable progress toward a bigger-brained future was blocked by what seemed an insurmountable obstacle. Either bipedalism—or brain growth—would have to be sacrificed.

However, the forces of evolution were still at work. Gradually, and unconsciously, human beings began to evolve a changed pattern of development. Mothers started giving birth to children whose brains were to do most of their growing outside of the womb—a process that has reached a point today at which the average human brain at the time of birth is only 25 percent of the size it will reach as an adult, compared with 65 percent for a chimpanzee. The advantage of this new arrangement lay in the fact that hominids' brains now quadrupled in size during their lifetimes. There was one disadvantage in this change, however, in that throughout the period that the growth of the brain was at its most intense—during the first few years of life—the hominids' body growth now became retarded.

The price that humankind paid for this increased brain capacity in later life, therefore, was an increased helplessness at the start of it. Whereas a newborn horse could stand up on its own and walk within two hours of birth, and a baboon could look after itself at twelve months, human babies now became utterly dependent on their mothers for the first six years of life.

This called for a completely new social structure. Previously, female primates had

The Taung skull, uncovered in South Africa in 1924 and named after the limestone quarry in which it was found, was the first firm evidence of the presence of early hominids in Africa. It belonged to an australopithecine child about six years of age. Raymond Dart, professor of anatomy at Witwatersrand University, quickly recognized that the skull must be hominid from its shape—too high and rounded for an ape—and from its full set of milk teeth, which are scarcely larger than those of a human child. Subsequent skeletal finds of *Australopithecus africanus* suggest that had the Taung child been male and grown to adulthood, it would have stood about four feet three inches tall and would have weighed from 100 to 110 pounds.

31

mated and then reared their children unassisted. Now the sexes began to cooperate in bringing up their helpless young, an arrangement that, in turn, required the setting-up of more stable and complex relationships between individuals.

Some distinguishing human characteristics can reasonably be supposed to have evolved as a result of—and as a response to—these social interactions. The reproductive cycle of the human female, which keeps women sexually receptive for most of the time, instead of just during the mating seasons, may well have developed both out of the community's need for parents to form stable bonds and out of the greater personal intimacy between male and female that was engendered by that stability; so may have the human female's ability to experience orgasm, unparalleled in any other species. It has also been suggested that sexual fidelity between men and women came about because their upright posture allowed them to mate face to face, the resulting eye contact having a bonding effect that was denied to those animals that were obliged to mate front to back. However, this theory is not supported by the example of the orangutan, an ape that mates face to face but is neither bipedal nor even approximately faithful.

Judging from the locations in which fossilized remains have been discovered, *Homo habilis* appears not to have strayed outside the boundaries of East Africa; archaeological finds in both France and Pakistan of what are claimed to be tools made by this hominid still remain unsubstantiated. But even though *Homo habilis* may not have felt prompted to find out more about the outside world, evolution did not stand still, and around 1.6 million BC, *Homo habilis* gave way to a more advanced race of hominids, who were known as *Homo erectus,* or Upright Man.

These new hominids were directly descended from *Homo habilis* but were built on a larger scale. Their biggest representative would have had as much as a twenty-inch height advantage and would have weighed forty-five pounds more than even the bulkiest *Homo habilis*. Possibly these new hominids had grown so much larger as an unconscious response to food shortages. For the bigger an animal gets, the lower its energy requirements become in proportion to the size of its body; a marmoset, for instance, needs proportionally three times as much energy as a human being and, therefore, has either to find supplies of high-energy food—such as nuts or meat—or to spend virtually the entire day eating low-energy foods, such as grass and leaves. Faced, therefore, with an environment in which high-energy food was scarce but low-energy materials abounded, hominids might well have evolved larger frames in order to accommodate this imbalance.

It was not only their bodies that had grown. The average *Homo erectus* brain now measured 950 cc, as opposed to *Homo habilis'* 800 cc. With this improved mental and physical capability, the new hominids had taken the art of toolmaking far beyond the mere chipping of rocks. From the right piece of stone, *Homo erectus* now had the ability to make symmetrical, pear-shape implements known as hand axes, with which they could more easily butcher whole carcasses of meat. They had also learned how to make stone scrapers—for jobs such as cleaning flesh off of hide—by knocking small flakes off a larger piece of rock.

At some point, they also discovered how to use fire. The first traces of this are equivocal: At Chesowanja in East Africa, researchers discovered pieces of burned clay, which were dated to 1.4 million BC. Animal bones and stone tools found nearby show that early hominids—probably *Homo erectus*—had been present, and tests on

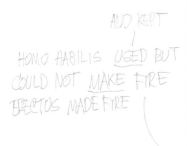

AND KEPT

*HOMO HABILIS USED BUT
COULD NOT MAKE FIRE
ERECTUS MADE FIRE*

the clay indicated that it had been heated to 750° F., a temperature typical of a campfire blaze. But it is difficult to be sure; other specialists point out that the clay could have been burned in a bushfire.

There is no definite evidence for the controlled use of fire by humans throughout the next million years. Caves at Zhoukoudian near Beijing have yielded human remains from about 460,000 years ago, along with layers of what was originally thought to be burned charcoal, but what has now been identified as burned owl and bat droppings. Again, it is not possible to ascertain whether these droppings had been burned deliberately by human beings, or whether, in fact, they had merely been consumed by a naturally occurring fire.

The exact date at which *Homo erectus* acquired control over fire remains unknown, therefore. What we are sure of, however, is that *Homo erectus*—unlike *Homo habilis*—lived not just in the balmy climate of Africa but also in the chillier zones of Europe and China. If, as seems probable, these hominids crossed over into these colder continents from their native Africa, fire would have been essential to them in securing a foothold. Without it, they certainly would never have been able to survive a winter on the exposed and windblown steppe of eastern Europe.

Not that the opening part of the Quaternary period, from two million to 8000 BC, was particularly hospitable anywhere in the world. During this time, periodic variations in the earth's orbit and constant shifts in the shape of the world's continents combined to produce a perpetually fluctuating climate, alternating approximately every hundred thousand years between ice ages and warmer intermediary periods known as interglacials.

During the interglacials, such as the one that the earth is enjoying today, the equatorial regions experienced an abundant rainfall, while the northern and southern parts of the world experienced relatively temperate climates. However, each interglacial gradually grew colder, and the ice spread southward until it covered large parts of Europe and North America. Northern Asia became bitterly cold, but lack of rain kept it free of ice. Then, after a few thousand or, sometimes, just a few hundred years, this glacial period reached its peak and quite suddenly was replaced by a new interglacial—and the cycle began again.

Throughout this alternation of climates, *Homo erectus* doggedly survived, changing remarkably little either physically or mentally over a period of approximately 1.5 million years. These hominids learned to make slightly improved tools, and they mastered the construction of twig huts, but their biggest strides forward came in hunting. By about 300,000 BC, bands of nomadic *Homo erectus* hunters were tackling—and killing—animals as large as elephants, and they were using fire to drive big game into enclosed areas where they could be slaughtered more easily.

They moved around in bands of twenty to thirty individuals; Africa, Asia, and Europe together might have supported 40,000 such groups—fewer than one million people across the whole globe. How they lived is a matter for conjecture. Probably the complex vocal articulations of human beings today were beyond them, and in any case, coordinating a band of hunters to bring down large prey can be done by gesture and example; wolves manage it quite efficiently.

And so, hominids continued, making their own slow, unspectacular way along the evolutionary path. However, soon after 300,000 BC, that path seems to have diverged. We know this from skeletal remains found all over Africa, Asia, and Europe that, around this time, start exhibiting a mixture of physical characteristics: They have

Shown above as it appeared nearly 2 million years ago, Olduvai was the site of a lake surrounded by lush vegetation that attracted a wealth of animal life. As streams running off nearby volcanic mountains slowly silted up the lake, hominid habitats were buried by sediment. Later, geological movement shifted the drainage basin to the east, and a stream began to carve a nine-mile-long gorge. The erosion of this now-arid landscape *(left)* uncovered clues to human evolution.

A CRADLE FOR EARLY LIFE

Through the dedicated work of a husband-and-wife team of archaeologists, the Olduvai Gorge in Tanzania has yielded a unique insight into the cultural and practical world of hominids between 2 million and 700,000 years ago. Louis Leakey, a pioneer of the search for human origins in Africa, was attracted to Olduvai by its abundance of fossils and animal bones, which led him to believe that it might also have been home to hominids. After spending almost thirty years uncovering thousands of fossils and stone artifacts, he and his wife, Mary, were rewarded in 1959 by the discovery of a 1.75-million-year-old hominid skull. They nicknamed this first fossil evidence of *Australopithecus boisei* their Dear Boy. Two years later, they discovered *Homo habilis*, the earliest-known *Homo* species, whom they credited with producing the older stone artifacts that they had found; hence, they called him Handy Man. The later tools found in the gorge are the work of *Homo erectus*. After Louis Leakey died in 1972, his work was continued in other parts of Africa by his wife and his son Richard.

Tools found at Olduvai include a primitive pebble chopper from 1.8 million years ago *(below, right)* and a more sophisticated 800,000-year-old hand ax *(below, left).*

Homo habilis had a longer skull than early hominids—as in this reconstruction of a male head—which made room for a larger brain.

the larger brains and thinner skulls of a modern *Homo,* but they retain the heavy bones and the flat, rather than high-domed, foreheads of an altogether more primitive creature. These remains were of hominids who came to be classified as archaic *Homo sapiens.* Some of these seem to have gone on to evolve naturally into anatomically modern man, or *Homo sapiens sapiens.* Others—particularly those found in Europe and the Levant—underwent a quite different kind of transition. They became, by 100,000 BC, Neanderthals, creatures whose name has since become—most unfairly, as it happens—a byword for the epitome of primitive brutishness. Scientists previously classified these beetle-browed hominids as an entirely separate species—*Homo neanderthalensis*—but they have now been firmly instated as our very close relative, *Homo sapiens neanderthalensis,* whose cultural and technological achievements are discussed in chapter two.

Compared to their predecessors, the Neanderthals were to hold only a limited tenure on the earth, for by 30,000 BC, they had all but disappeared. Whereas *Homo erectus* had existed for more than one million years and had made remarkably few innovations, the Neanderthals vanished after just 70,000 years, during which time they had, among their many other achievements, revolutionized the art of toolmaking to the point where it could be classed as mass production.

The creatures that replaced them were startlingly different in appearance. Although less than eight inches taller than the average Neanderthal, they were infinitely more slender and nimble and weighed from sixty-five to 100 pounds less. There were changes in their features, too: Above their eyes rose tall, elegant foreheads—uninterrupted by heavy Neanderthal brows—that continued upward at first and then sloped back downward, forming the domed cranium that housed the brain. This was *Homo sapiens sapiens,* Man the Double Wise, a human being who was anatomically indistinguishable from modern humans. These creatures first appeared in South Africa around 100,000 BC and, within the next 70,000 years, had replaced all the previous species of hominid throughout the entire globe.

Precisely how they achieved this remains one of the enduring mysteries of the human story. That *Homo sapiens sapiens* came to dominate the rest of the world with their technology and culture is unquestioned, but what no one knows is how they achieved this in what is, in biological terms, a minuscule amount of time.

Possibly, it was because they had at their disposal a tool that was to prove more useful than any of their bone sewing needles, a weapon more powerful than any of their stone spearheads. That tool was language.

With language, *Homo sapiens sapiens* could pass on technological knowledge. With language, they could mastermind complex cooperative ventures, such as what appears to be an organized seaborne migration to Australia in about 50,000 BC. And with language, they could give expression to abstract thoughts and concepts that were manifested in the great outpouring of paintings, drawings, and carvings that began in Europe about 30,000 BC.

They unquestionably had all the vocal equipment for it. Their pharynges had grown proportionally much longer than those of earlier hominids, and their tongues had an increased flexibility that enabled them to form and emit a much

Squatting in silent contemplation, this *Homo erectus* male *(right)* grips a pear-shape hand ax in strong, fine-boned fingers. *Homo erectus,* generally regarded as the precursor of modern *Homo sapiens sapiens,* had limbs and torso similar to a human of today. The pronounced brow ridges, deep-set eyes, and protruding jaw, however, proclaim his descent from *Homo habilis,* the earliest toolmaker. The ax—probably used for chopping, cutting, and pounding food—is one of a range of simple but versatile tools that appeared soon after the emergence of *Homo erectus* about 1.6 million years ago.

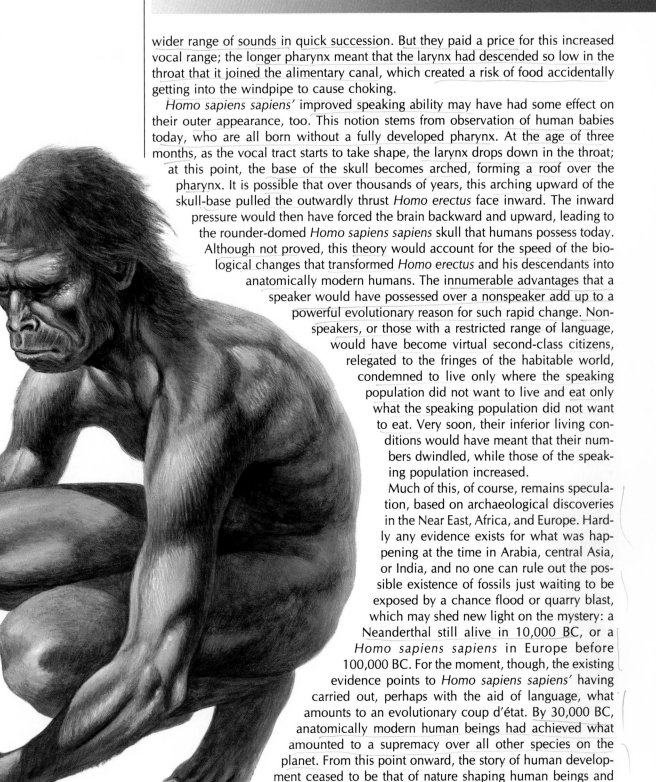

wider range of sounds in quick succession. But they paid a price for this increased vocal range; the longer pharynx meant that the larynx had descended so low in the throat that it joined the alimentary canal, which created a risk of food accidentally getting into the windpipe to cause choking.

Homo sapiens sapiens' improved speaking ability may have had some effect on their outer appearance, too. This notion stems from observation of human babies today, who are all born without a fully developed pharynx. At the age of three months, as the vocal tract starts to take shape, the larynx drops down in the throat; at this point, the base of the skull becomes arched, forming a roof over the pharynx. It is possible that over thousands of years, this arching upward of the skull-base pulled the outwardly thrust *Homo erectus* face inward. The inward pressure would then have forced the brain backward and upward, leading to the rounder-domed *Homo sapiens sapiens* skull that humans possess today. Although not proved, this theory would account for the speed of the biological changes that transformed *Homo erectus* and his descendants into anatomically modern humans. The innumerable advantages that a speaker would have possessed over a nonspeaker add up to a powerful evolutionary reason for such rapid change. Nonspeakers, or those with a restricted range of language, would have become virtual second-class citizens, relegated to the fringes of the habitable world, condemned to live only where the speaking population did not want to live and eat only what the speaking population did not want to eat. Very soon, their inferior living conditions would have meant that their numbers dwindled, while those of the speaking population increased.

Much of this, of course, remains speculation, based on archaeological discoveries in the Near East, Africa, and Europe. Hardly any evidence exists for what was happening at the time in Arabia, central Asia, or India, and no one can rule out the possible existence of fossils just waiting to be exposed by a chance flood or quarry blast, which may shed new light on the mystery: a Neanderthal still alive in 10,000 BC, or a *Homo sapiens sapiens* in Europe before 100,000 BC. For the moment, though, the existing evidence points to *Homo sapiens sapiens'* having carried out, perhaps with the aid of language, what amounts to an evolutionary coup d'état. By 30,000 BC, anatomically modern human beings had achieved what amounted to a supremacy over all other species on the planet. From this point onward, the story of human development ceased to be that of nature shaping human beings and began to be that of human beings shaping nature.

LANDSCAPES OF THE ICE AGE

When the last Ice Age reached its zenith around 16,000 BC, average temperatures in parts of Europe and North America were as much as 65° F. lower than they are now. Between one-third and one-quarter of the entire land surface of the globe lay entombed beneath a shroud of ice, in some places almost two miles deep.

Much of this mass was in constant motion. In some areas, cliffs of frozen water more than 325 feet high formed the spearhead of advancing ice sheets that progressed at a rate of 2,000 to 4,000 feet a year, gouging and tearing up the ground as they went, either grinding huge boulders into powder or picking them up like marbles and dumping them hundreds of miles from their source.

In the Northern Hemisphere, the ice traveled south from the Arctic Circle, thrusting out in a number of different directions. North America had two giant ice sheets: One came down the East Coast, blotting out eastern Canada and New England, then heading inland across the Midwest until it eventually reached the point where Saint Louis, Missouri, stands today, more than 275 miles southwest of Chicago. The second ice sheet extended down the West Coast from the Rockies, engulfing parts of Alaska, all of western Canada, and parts of Washington State, Idaho, and Montana. At the same time, an ice sheet in Europe stretched out from Scandinavia, swallowing up most of the British Isles and Denmark and large parts of northern Germany, Poland, and the Soviet Union.

Because the Southern Hemisphere consisted largely of open ocean, less land was lost to the rapacious invader. In tropical Africa, some glaciers formed on high ground, but only the Andes in South America and the mountains of New Zealand saw this process reach spectacular proportions. Meanwhile, the Antarctic ice sheet thickened and spread north.

Because so much of the earth's water was locked up in ice, dramatic changes occurred in the planet's geography and climate. Sea level dropped by 410 feet, exposing vast areas of what had once been seabed. The English Channel and the Persian Gulf disappeared; in Southeast Asia, a huge new continent emerged, linking the islands of Indonesia in one large landmass; Siberia and Alaska were united by a broad land bridge. Rainfall levels dropped, too; average rainfall in Britain was the same as that of present-day Morocco. Fierce, strong winds blew, scattering huge quantities of dust, rock debris, and vegetation.

The pressure of the ice, coupled with the increased hostility of the climate, forced the earth's vegetation belts much closer to the equator than they are today, as demonstrated by the map on the right. At latitudes where dense forests now flourish, arctic tundra stretched out bleak and featureless; in similar fashion, regions that today are deciduous woodland were, in 16,000 BC, treeless, grassy steppe.

Four of the most common environments prevalent at this time are shown on the following pages. In each, human beings and other animals were compelled to adapt their lifestyles to the resources available. In the process, they developed strategies that enabled them to weather the full force of the Ice Age in all its varied and daunting manifestations.

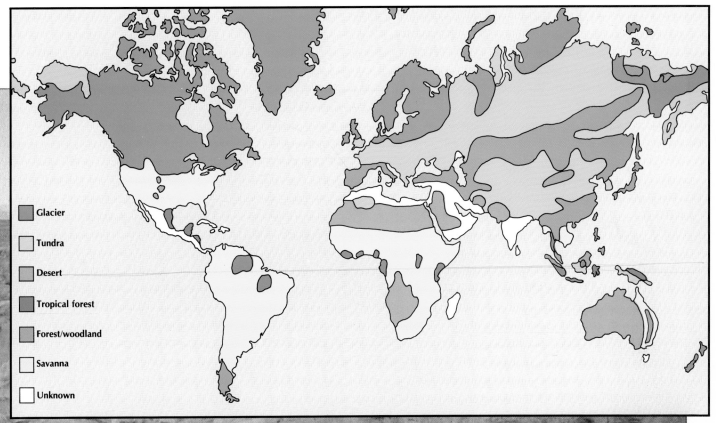

Glacier
Tundra
Desert
Tropical forest
Forest/woodland
Savanna
Unknown

THE CHEERLESS TUNDRA

At their southern extremities, the great northern ice sheets yielded to an environment of tundra—a flat, semifrozen zone. In North America, the area of tundra was smaller than it is today, but in Europe and Asia, it stretched from the Atlantic coast of France right through southern Germany, eastern Europe, and Siberia, to China.

Fiercely cold winters melted into short, cool summers, when mosses, heathers, and arctic flowers briefly carpeted the great plains. However, no tree could take root, because the tundra soil remained permanently frozen just beneath the surface.

Farther south, the tundra merged with warmer and grassier steppe, to which large herbivores such as bison, mammoths, and the giant deer—with an antler span of ten feet—migrated in winter, along with horses and the two-horned woolly rhinoceros.

Most human beings did not venture deep into the tundra outside of the summer months. However, the hardier—or hungrier—hunters did brave the howling winter winds to go in search of those animals that could survive the cold, among them the shaggy muskox and the elusive Arctic wolf.

Muskox

Woolly rhinoceros, now extinct

Collared lemming

Giant deer, now extinct

FRINGES OF THE FOREST

During this period, no woodland grew north of the Mediterranean Sea. To the south, however, oak, beech, walnut, and chestnut trees spread a graceful canopy across a broad sweep of the world that stretched from the southeastern part of North America, through southern Europe, to southern China and Japan.

Few human beings chose to live in the forest itself, preferring to settle at the edge of a wooded area, ideally near a river or the ocean. The trees provided a rich supply of fruit and berries in season as well as fuel and materials for shelter all year round. Caves offered a refuge for both humans and other hungry predators, which in the woodland environment of southern China included tigers, black bears, and hyenas. Other large but strictly herbivorous occupants of the woods were the giant panda and the stegodon, an elephant-like creature with tusks so close together its trunk could not fit between them.

Those hunters who did go into the forest brought back wild pigs and small deer such as the muntjac. They also gathered chestnuts, walnuts, and any other fruit and seeds that had escaped the attention of the energetic macaque monkeys.

Stegodon, now extinct

Giant panda

Spotted hyenas

Macaque

THE ARID ZONES

Paradoxically, hot deserts flourished during the Ice Age, because the cold weather inhibited evaporation—and, therefore, rainfall. Temperatures near the equator remained high, and the largest arid zones were found in the Sahara and Kalahari areas of Africa, as well as in Arabia, northwest India, and Australia.

In the most parched regions, desert dwellers could live only in those places where they had access to water in the form of permanent natural springs, water holes that filled up during periods of heavy rain, or seasonal streams that were fed by melting glaciers in nearby mountain ranges. Human beings survived, as did the desert wolves and jackals, by hunting the herds of gazelles and wild asses that migrated through their territory.

In less-barren territory, often close to springs or marshes, enough grass grew to sustain the large, grazing aurochs, or wild oxen. Humans, too, could usually find enough grass seeds, roots, and tender plant stems to keep them alive. Occasionally, they might even manage to supplement their diet with some protein in the form of ostrich eggs or even the flesh of a small animal such as a tortoise or a Cape hare.

Aurochs, now extinct

Gazelles

Ostrich

Tortoise

Cape hare

Sugar glider

Koala bears

Extinct *Thylacoleo* with young in its pouch.

Lakes were highly prized by the people of the last Ice Age. They guaranteed a supply not only of fresh water, fish, and crustaceans, but also of the animals and water birds that gravitated naturally to the lake and could be killed for their meat while they bathed or drank.

Around 16,000 BC, however, a global drought was drying up these desirable lakeside habitats. The process had reached an advanced stage in Australia, presenting problems both for human beings and for the animals with which they shared the continent. Many of these creatures formed a part of the human diet: hopping mice, wallabies, kangaroos, koala bears, the echidna, or spiny anteater, and the sugar glider, a flying opossum that carried its young in a pouch. Humans also hunted larger beasts, including a ten-foot-tall kangaroo named *Procoptodon* and a distant, rhinoceros-size relative of the burrowing wombat, known as *Diprotodon.* In turn, humans would have been wary of the stalking *Thylacoleo,* a ferocious marsupial lion about the size of a modern-day leopard, which possessed massive, fearsome jaws.

Echidna

Giant kangaroo, or *Procoptodon*

THE PEOPLING OF THE EARTH

On a chilly summer afternoon, beneath a watery sun, a small band of some thirty people—men, women, and children—might have been seen advancing slowly across the cheerless tundra of eastern Siberia. The men, armed with fearsome, stone-tipped spears and protected against a raw wind by tightfitting clothing made of animal hide, walked at the head of the party. As they moved, they kept a nervous lookout for bears and other predators, noting any likely hunting spots they passed, and scanning the ground for droppings or hoofprints indicating that reindeer or mammoths had passed that way.

Toward the back of the group came the women, picking their way carefully around the low bushes and lichen-encrusted rocks that littered the ground and slogging through sodden moss and grass. They stooped as they walked, bowed under the weight of carrying small children and the rolled-up lengths of fur and hide that they would use to erect shelters for the night. Older children ran and played around the group as it progressed slowly across the plain toward the higher ground that could be seen ahead, stretching away to the range of low hills on the distant horizon. Like the men, the women and children kept an eye on the terrain that they were crossing, ever alert for anything it might have to offer in the way of food—seeds, nuts, and berries, as well as animals.

We cannot say with certainty who these people were. We do know, however, that between 50,000 and 13,000 BC, they—or members of a group that was very much like this one—made this journey during a period in the world's development when Asian Siberia was joined to American Alaska by a land bridge approximately fifty miles long. At one point in their journey across this bridge, the travelers would have stepped over a nondescript patch of scrub or stones that could be said to constitute the dividing line between the two continents. In so doing, they became the first human beings ever to set foot in America.

Not, of course, that our imagined band had any concept of such momentous implications. As far as the members of this group were concerned, they were merely progressing a few more weary and unremarkable miles across the seemingly endless and windy steppe, in search of a new homeland. It had been the gradual depletion of food resources, as well as the increasing competition from bands of eastward-traveling migrants that had made them leave their own ancestral hunting grounds in Siberia and go eastward as well, taking advantage of the brief weeks of mild summer weather to make their move.

They had always been a nomadic people, shifting regularly from camp to camp across their old territory and making the best use of all the different resources it offered at different times of the year. They were accustomed to traveling light, and they carried with them tools that were simple and disposable, such as rough stone

The face of *Homo sapiens sapiens,* clearly modern in form, stares from beneath sharply engraved lines representing braided hair. Excavated at Brassempouy in France, this tiny head—measuring less than one and one-half inches high—was carved from the ivory tusk of a great woolly mammoth at some point between 30,000 and 20,000 BC. By this time, early humans had spread out of Africa, across the temperate zones of the globe, and were developing artistic and technological skills to match their territorial expansion.

The main sites at which fossilized bones of *Homo erectus* and other remains, such as artifacts and butchered animal carcasses, have been found are shown on the map above. As the third identifiable hominid, *Homo erectus* marked an intermediate stage between the more apelike *Homo habilis* and the modern human, *Homo sapiens sapiens*. Emerging in East Africa some 1.6 million years ago, the species had reached Java by 800,000 BC. In Europe, the earliest skeletal traces—such as the Arago skull and the fossilized Heidelberg jaw—date from around 400,000 BC, but remains of human artifacts indicate a presence soon after one million BC. The mastery of fire brought safety from predators and promoted new forms of cooperative social organization, thus speeding the expansion of *Homo erectus* across Eurasia.

flakes that they could quickly shape and sharpen by skilled chipping. Their prized possessions were similarly few: a handful of figurines, along with small ornaments made of bone, ivory, or stone. Even their more permanent implements—delicately wrought bone needles, sharp flint blades set in bone or wood handles, and finely honed borers for piercing hide and engraving bone or horn—could all be bundled quickly into bark containers or hide shoulder bags when the time came for striking camp and traveling onward.

Now, the gently rising land did indeed seem to be growing richer. In the distance, reindeer grazed. Fanning out and sweeping around to approach the herd from downwind, two of the men, accompanied by an adolescent boy learning hunting skills, silently stalked a young buck, which was feeding apart from its companions. The animal, standing more than three feet tall at the shoulder and weighing more than the three hunters combined, continued to graze unsuspiciously. Stealthily, one of the men fitted a spear into the wooden throwing stick that rested on his shoulder. He paused for a moment, and then, with a sharp flipping action, he dispatched the spear. His aim was true; the buck pitched forward, wounded. Still wary, the other men moved in on the crippled animal, mindful of the flailing hoofs and of the lethal antlers. Again and again, they darted in to stab the dying beast at close quarters, until, at last, it lay still.

Quickly now, as dusk was drawing near, they set about the task of butchering the animal. To accomplish this, they used a mixture of implements—smooth and sharp flint blades for skinning, and much heavier stone choppers for smashing the thicker bones and more resistant sinews.

When their work was done, the three rejoined the main body of the band, their shoulder pouches full to bursting with fresh meat. That night, beside the newly constructed hide shelters, reindeer steaks sizzled over the campfire, and well-fed children slept soundly.

These first Americans undertook their journey toward the end of a two-million-year-long period during the course of which both human beings and the world they inhabited experienced a series of great climatic shifts—long, protracted freezes and sudden thaws. Enacted over tens of thousands of years, these changes were sometimes slow, sometimes perceptible within the span of one human life; but over a number of generations, they dramatically altered whole environments, locking seas and rivers in great sheets of ice that later thawed, increasing sea levels and submerging large tracts of land.

In addition to geographical schisms, the early Quaternary witnessed the opening of a definitive genetic gap between humans and apes. This would leave the bipedal australopithecines firmly on the ape side, where they would die out. It also established *Homo habilis*—Handy Man—as the first indisputable members of the genus *Homo,* to be joined later by their larger and more sophisticated counterparts: *Homo erectus* (Man the Upright), *Homo sapiens* (Man the Wise), and finally, the anatomically modern human, *Homo sapiens sapiens* (Man the Double Wise), who, in evolutionary terms, is indistinguishable from people today.

The physical difference between *Homo habilis*—100 pounds of ape-jawed primitive—and *Homo sapiens sapiens*—who, dressed in modern clothing, would easily pass for a 160-pound sales executive—was immense. Although more advanced genetically, *Homo sapiens sapiens,* in cultural and technological terms, inherited a great deal from their less-refined ancestors. It was they who had acquired fire, who had devised ever more intricate methods of toolmaking, and who had, over thousands of generations, evolved increasingly complex patterns of work and social life, thereby giving the basically weak and defenseless human species a distinct competitive edge over all other animals. These skills ultimately would enable humans not only to gain mastery over rival species and to cope with environmental and climatic changes but also to colonize the entire globe, from the blazing hot equatorial tropics to the frozen wasteland of northern Siberia.

From the very beginning, as far back as 2.3 million BC, *Homo habilis* had developed a taste for meat. It was a food that had appeal in terms of nutrition as well as flavor, because, compared with fruit and vegetables, meat is a much more efficient way of taking in nourishment. Two pounds of venison, for instance, provides almost six times as many calories as the same weight in berries or green vegetables. Thus, the securing of one deer provided the same amount of nourishment as many hours of backbreaking foraging on the hillsides and scrublands. In addition, meat eating gradually destroys the organisms required by the human digestive system to break down the cellulose in vegetables, which then become more difficult to digest. As a consequence, over a number of generations, this process must have contributed to humans' following a more carnivorous diet.

Nevertheless, killing large wild animals was not an easy task, and as powerful as the first hominids' cravings for meat might have been, they often must have gone unsatisfied. Australopithecines, for example, had been neither swift nor sharp enough, nor had they developed sufficiently advanced tactics, to hunt game successfully. *Homo habilis* was better equipped physically, with a far more developed manual dexterity and a brain that was almost 50 percent larger. With these assets, *Homo habilis* developed ways of stalking and killing small animals (as troops of present-day baboons, working cooperatively, have been observed to do) and of driving large animals into swamps or over cliffs, where they could be dispatched with a hail of rocks.

Since rudimentary hunting was a highly opportunistic, unsystematic affair, however, *Homo habilis* and their australopithecine predecessors still relied on scavenging as the primary means of acquiring meat with any degree of regularity. What this meant was that, apart from those occasions when they had the good fortune to surprise their prey or to come across a weak or injured beast, *Homo habilis* hunters were obliged to scavenge what meat they could from the carcasses of animals that either had been left by larger carnivores or had died of natural causes. Often, especially when they were hungry, the early hominids must have had to drive other feasting animals, such as lions and leopards, away from the carcass, a hazardous

AN EARLY EDEN

Shown as it appeared around 400,000 years ago, this site of a hunter-gatherer encampment *(right)*, sheltered in a deep river valley at Bilzingsleben in what is now Germany, commands a richly diverse environment. As the diagram at far right demonstrates, each type of vegetation was home to a variety of animal and plant species, which together satisfied all the material needs of the human community that settled in this place. The rhinoceros, for example, the most commonly hunted animal, provided not only plentiful meat and bone marrow but also large bones for constructing shelters and for making tools. The fur of beaver, deer, and other animals was used to make clothing.

business if one is little more than four feet tall and armed with nothing more substantial than a stick, or, at best, a stone tool. This being the case, their diet must have remained largely vegetarian, although they may have been able to supplement it—as chimpanzees do—with termites or other insects that they worked out of their nests, using long straws or twigs.

As time wore on, however, the later hominids changed physically, and as they changed, they developed an increasing capacity to alter the world around them. They might have lost the ability of tree-dwelling primates to grip with their feet, but this was more than compensated for by the fact that their hands had become ever more sophisticated instruments. Whereas all primates could seize objects with their hands and let them go at will, only the hominids developed the distinctive opposable thumb, which could allow, by pressing back against the other gripping fingers, the delicate handling of objects and an infinite number of precise digital adjustments. In this respect, *Homo erectus'* hands scarcely differed from our own, and this manual

Large herds of bison, deer, and horses roamed wide swaths of open pasture.

Deciduous woodland sheltered elephants, aurochs, wild pigs, and deer, and it also yielded an abundance of nuts, berries, herbs, and other edible plants.

Scrub was home to rhinoceroses and elephants; roots, berries, and nuts were also plentiful there.

Marshes were inhabited by water birds and waders, nesting among the reeds and stalks.

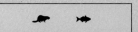

Rivers, streams, and pools were stocked with a variety of freshwater fish; colonies of beaver lived on the edges of land and water.

On land, bears and predators—lions, wolves, and foxes—roamed from one type of vegetation to another.

dexterity allowed later hominids to produce stone chopping and cutting tools that surpassed the crude creations of other primates.

In toolmaking, the most distinctive product of *Homo erectus* was the hand ax. These pear-shape tools of rounded triangular profile, with a point at the top and a broad curved butt, remained one of the most important tool types in Europe, Africa, and western Asia from 1.5 million BC up to about 100,000 BC. Why they continued to be used for so many millennia is still something of a mystery, but they clearly demonstrate that *Homo erectus* and their immediate successors were conservative in their tool manufacturing and were content to go on creating similar shapes of stone tools for hundreds of thousands of years.

Another major contribution to technological progress was *Homo erectus'* mastery of fire. Scientific opinion varies as to how this was achieved. Some say that hominids first created fire by accident from sparks produced by rubbing together sticks or stones. Others say that they literally captured the flames by dipping sticks into the edge of either forest fires or volcanic lava flows.

Whatever the truth of the matter, there can be no doubt that fire immeasurably improved the quality of hominid life. It provided warmth, of course—increasingly important outside the tropics as the ice ages progressed. Indeed, it is very likely that the colonization of Europe and Asia by *Homo erectus* that began around a million years ago was made possible only by the knowledge of fire, both as a tool to ward off predators and as a means of keeping warm in the fierce cold of northern winters. Other species had to depend on genetic changes—such as the development of a warmer coat—to adapt to colder climates, but humans could rely on their intelligence and ability to make clothes, fire, and shelter.

Fire could also be used for cooking meat. Cooking constituted a significant break-through in more ways than one. It made meat not only more appetizing but also more digestible and beneficial, because heat breaks down some of meat's more complex compounds and releases a variety of nutritious juices. Cooking meant that *Homo erectus'* teeth grew smaller, since they were no longer required to bite through large chunks of raw flesh; and the introduction of softer, cooked food also removed the necessity for the large, obtrusive jaws that humans had needed to assist them in what must often have been the epic task of chewing raw meat. Some indication of what mealtimes must have been like 1.5 million years ago can be gleaned from a *Homo erectus* jawbone—found near Heidelberg, Germany—which has cut marks on the front teeth. Evidently, *Homo erectus* ate meat by clamping it between the teeth and then hacking at it with a sharp flint tool.

However, the most fundamental contribution that fire made to improving the lot of human beings was its capacity to inspire fear in other animals. For, while the hom-inids had over millions of years become meat-eating predators themselves, they had not developed anything like the natural defenses that were needed to fight off rival carnivores such as lions or saber-toothed cats. Such predators could wreak havoc among the hominid groups, and bands venturing from their shelter at night were likely to be torn to pieces by predators that had picked up their scent. Skeletal remains found in South African caves give graphic testimony to this vulnerability. In a cave at Swartkrans, halfway between Johannesburg and Pretoria, the skull of one unfor-tunate australopithecine, dating from 1.5 million BC, was found, scarred by the deep and unmistakable toothmarks of a leopard. The big cat apparently took its prey into a tree to eat, out of the reach of competitors, and, at the end of the meal, dropped the discarded skull into a nearby cave shaft.

Fire, however, kept all such predators at a safe distance. The importance that early humans attached to the flame is underlined by the fact that some groups of hominids kept their fires burning—or at least smoldering—constantly, day and night, both inside and outside their caves or shelters.

The fire thus became the center of the community's activities, the focus for a far more complex social life than had yet been seen. The idea of a home base, around which a community could organize its life more cooperatively and efficiently than when permanently on the move, was taking root for the first time.

Along with the concept of the home base went the development of shelters. As members of the *Homo erectus* species spread from their tropical homeland to colder regions in Europe and Asia, they relied on their cultural skills to adapt to the harsher climate. The discovery of fire was a major asset, but without clothing or shelter, the colonization of the temperate latitudes would have been impossible. In some areas, caves provided convenient natural shelters; but where these were lacking, *Homo erectus* had to improvise with the materials on hand to build shelters. One of the earliest is the campsite of Terra Amata, on the Mediterranean coast of France, near the modern-day city of Nice.

Here, archaeologists found the remains of eleven oval huts, each constructed of branches resting on a central row of upright posts. Around the edge of each hut, the bottoms of the branches were held in place by heavy stones. Near the center of the hut was a hearth, and the many other traces found provide a thumbnail sketch of life in this early European settlement. The occupants sat and slept on animal skins and may have painted their bodies with ocher. A faint trace in the sand showed where a wooden bowl had been placed. A scatter of flint chips reflected the efforts of a toolmaker. From bone remains, it is known that the people of Terra Amata ate red deer, elephant, rhinoceros, goat, and boar; and shells of oysters, mussels, and limpets show that the seashore was not far away.

As part of their social development, members of the *Homo erectus* species probably took further than their predecessors the sexual division of labor, which had evolved along with the evolving human brain. As hominids had developed bigger and bigger brains, so it had become biologically necessary for baby hominids to be born at earlier—and more helpless—stages in their growth.

The demands of constant childcare restricted the movement of women, and the presence of clinging, chattering youngsters certainly hindered them in the stalking of easily alarmed prey. So, the task of ranging far and wide, scavenging and hunting for meat, devolved upon the males, while that of foraging closer to home for vegetable food and gathering it in became the province of the females. The hunter-gatherer lifestyle, as it is known, survived for almost one million years, during which time, *Homo erectus* became firmly established outside Africa, across Europe and Asia, and in China, Java, and other parts of the Far East. This way of life continued for so long that it was even inherited by the different subspecies of *Homo sapiens* to which *Homo erectus* gave rise.

Of *Homo sapiens* subspecies, the most famous representatives today are the Neanderthals, so called because the first remains of one of their number were found in 1856 in Germany's Neander Valley, near Düsseldorf, when workmen at a limestone

quarry dug up a skullcap, ribs, and fragments of pelvis and the bones of limbs. The owner of the quarry thought the bones belonged to a bear. The local schoolteacher was of the opinion that they belonged to someone who had been washed into the quarry during the Biblical Flood. A German anatomist confidently suggested that the bones were those of a Russian Cossack who had fought against Napoleon in 1814 and, after deserting, had crawled into the cave and died. For almost a century, the word *Neanderthal* carried connotations of clumsiness and stupidity, based on the appearance of these bones. The myth was reinforced by the discovery, fifty years later, of the first complete Neanderthal skeleton—at La Chapelle-aux-Saints in southwestern France—which appeared to indicate that Neanderthals had had a stooping posture and slouching walk. Subsequently, it has been discovered that the shambling appearance of the man from La Chapelle-aux-Saints was not because of natural awkwardness but because he had suffered from arthritis.

However, many people had avoided the idea that the Neanderthals could possibly be our direct ancestors. These creatures presented an outer appearance that did not harmonize with the popular image of civilized human beings. Above their eyes spread an overhanging browridge of solid bone, topped by a long, flat skullcap that broadened into a bulging, bunlike cranium at the back and sides. The lower half of their faces thrust out dramatically from their necks and contained large front teeth, which they put to a variety of practical purposes: chewing animal hide in order to soften it, or clamping pieces of wood so that their hands would be free to carve or chop them. Their broad, flattened noses provided large nasal cavities in which inhaled air was warmed before entering the lungs, while their thrust-forward face structure kept this cold air well away from the brain. Stubby feet, short legs, and a massive, barrel-chested frame weighing anywhere up to 225 pounds completed the picture of a musclebound throwback.

Physically, they made both their hominid predecessors and their modern human successors look like weaklings. Their bones were more massive—some 10 to 20 percent heavier—than those of their modern successors, and the great muscles required to control them produced additional heat. For the whole of their history, between 100,000 and 30,000 BC, the Neanderthals experienced all the freezing rigors of a full-blown glacial climate. That is one reason that they developed extremely sturdy, robust frames.

The maintainance of body warmth was not, of course, the only reason for the

Contrasting forms of shelter demonstrate the ways in which early humans adapted to different climatic environments. Between 460,000 and 230,000 BC at Zhoukoudian in China, generations of *Homo erectus* lived permanently in deep limestone caves such as the one shown above, warmed by fires that also provided heat for cooking and security from wild beasts. In the gentler climate of southern France, their contemporaries at Terra Amata made do during summer hunting trips with light, temporary shelters built from leafy branches strewn over a rough framework of saplings (*opposite*).

Neanderthals' mighty physique. Wherever they found themselves, even in such temperate regions as the Middle East, the Neanderthals led a tough existence. Their nomadic way of life was a punishing regimen that involved walking twenty miles a day over rough and frequently icebound terrain, often while carrying heavy loads. Upper body strength was essential, and the Neanderthals' handgrip was probably two or three times as powerful as that of a modern human.

And yet, for all their rugged, brawny appearance, the Neanderthals were the match of later humans when it came to brain capacity. Whether this meant they could match them in intelligence is not certain, but what does emerge clearly from excavations of Neanderthal sites is that they took the art of toolmaking to new levels of refinement. Before the emergence of the Neanderthals, a tool made by any hominid—even the talented *Homo erectus*—had been a shaped stone, a rock painstakingly whittled away to produce an often-cumbersome implement. The Neanderthal workers, however, devised a means of producing two or three neat and virtually identical flakes from a piece of rock that previously would have provided only one. Using a technique called the "prepared core" method, the artisans first knocked the top off a round rock, much like slicing off the top of a boiled egg. This created a flat, disk-shaped surface, or core. The Neanderthals then worked around the edge of this disk, striking downward at the core with another rock to produce a series of flakes, which could be fashioned into knives or hand axes.

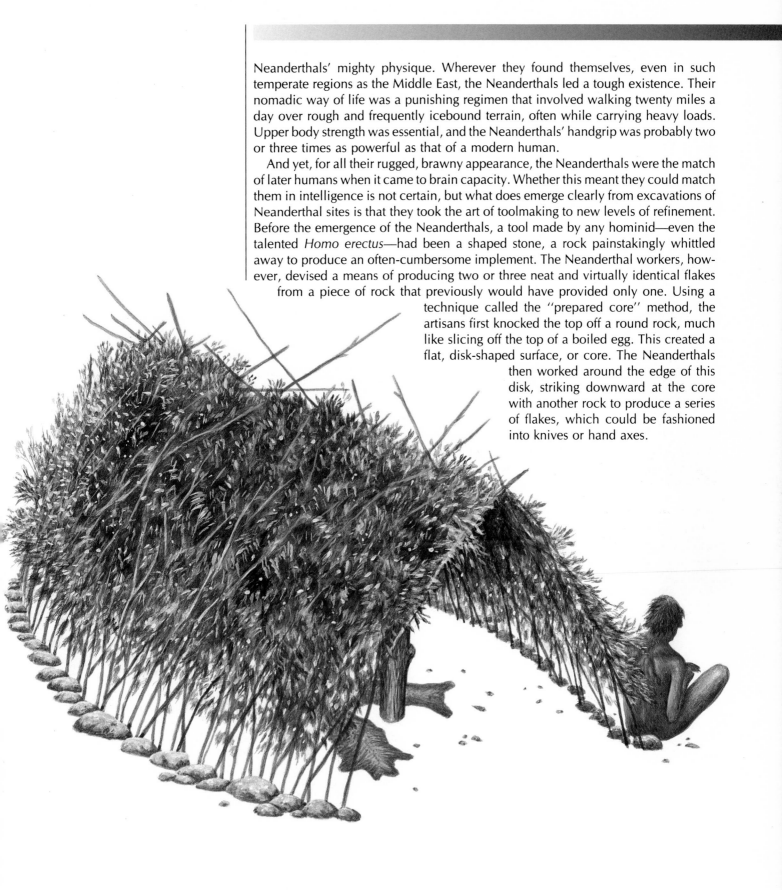

In addition to introducing this small—but nonetheless crucial—degree of mass production to the manufacturing of tools, the Neanderthals brought greater efficiency to the art of securing meat. Large-scale, preplanned game hunts became a regular feature of the Neanderthal era: Herds of horses, mammoths, and other large mammals were driven into muddy, swampy areas, where they could be more safely attacked at close range. Large quantities of meat—beyond the wildest dreams of the earlier hominids—could be obtained in this way. Whatever was surplus to immediate requirements would be stored in pits dug in the ground. Because permafrost began less than three feet below the surface in many areas of Europe at the time, meat placed in these natural refrigerators would stay fresh almost indefinitely. Bulk consumption of meat did not necessarily make for any lack of refinement, however; small scratches found on the teeth of Neanderthal skulls indicate that these giants probably used toothpicks after a meal.

The cultural life of the Neanderthals also outshone that of their ancestors. Some groups seem to have honored particular species among the animals they knew, and some made a practice of preserving in special shrines the skulls and bones of those creatures they killed. The cult of the bear, for example, seems to have been particularly widespread: One cave in Switzerland was found to contain thirteen carefully arranged skulls, some packed inside a rough chest made of heaped-up stones, others set in niches in the wall. Archaeologists excavating a French Neanderthal site unearthed, beneath a stone slab weighing more than one ton, a rectangular pit containing the bones of about twenty bears.

Little is known about what actually happened at such shrines. A cave in Lebanon contains evidence of a fallow deer having been ritually dismembered and sprinkled with red ocher, presumably symbolizing blood, but for the most part, it is possible

In an illustration based upon evidence discovered in Wyoming, spear-toting hunters of around 8000 BC close in on a herd of bison that they have backed up against a sand dune. This hunting technique was a refinement of the early hominid ploy of chasing prey into terrain that formed a natural trap. Once penned in, herds of animals started to panic, making it easier for the hunters, operating in groups of two or three, to capture a beast that they would not have dared confront under normal circumstances.

only to speculate on the precise nature and function of Neanderthal ceremonies. What can be said with some certainty, however, is that the lives of Neanderthal people contained a religious content.

Many people suspect, for example, that the Neanderthals believed in some form of afterlife. The food, tools, and animal skulls buried with Neanderthal corpses hint at a widespread belief in a world beyond the grave, where these items would either prove useful or provide some kind of magical protection. At La Ferrassie, in the Dordogne region of France, what appears to be a family cemetery has been excavated. Here, several bodies were found lying side by side: a man, a woman, a child approximately five years old, and two babies, all carefully arranged in shallow, parallel trenches oriented east to west. Flakes of flint had been buried with the man, and a flat tablet of stone was placed over his head and shoulders, either to protect him or to prevent his ghost from leaving the grave. Nearby, the skull of a child was found buried beneath a stone slab that was covered with strange markings; it was situated just a few yards away from the headless skeleton of a child—possibly the same individual. Near these graves, the excavators found a group of low mounds. Most of the mounds were empty, but one contained the skeleton of a baby, and it is thought likely that the others were also originally used for the remains of young infants, which are now completely decayed.

Further evidence of rituals involving dead children has emerged in a cave at Teshik-Tash, in the Soviet Union, close to the Afghanistan border. Here, the remains of a young Neanderthal boy have been found, his body laid out carefully amid ibex bones, with the horns of six ibexes neatly arranged in pairs to form a circle around his head. The ibex bones may be an indication that meat was buried along with the boy in order to sustain him on his journey to the next world. Again, the purpose and meaning of the rituals remain obscure; however, the combined evidence testifies to a growing fondness for ritual throughout Neanderthal times, which, in turn, suggests the evolution of a richer, more complex social life.

The most touching evidence of Neanderthal sensibilities was unearthed in 1957, in a large cave at Shanidar in the mountains of northern Iraq. Here, archaeologists uncovered the skeleton of a Neanderthal man who had died some 45,000 years earlier. His bones told the story of a life dogged by misfortune. His right arm was withered, probably as a result of a childhood injury. A smashed eye socket indicated that he was blind in one eye; and in addition to suffering from arthritis, he had sustained injuries to his legs, which must have made walking, let alone hunting, extremely difficult. Nevertheless, he not only survived all these handicaps, but he managed to go on living until he was forty years old—a ripe old age for Neanderthals, the oldest of whom lived to only about fifty. In the end, it took one final and spectacular piece of bad luck—a huge rockfall inside his own cave—to crush the life out of him once and for all.

That the Old Man of Shanidar, as he is popularly known, was able to stay alive for four decades is testimony to the care he must have received from his fellow Neanderthals. In a more primitive society, his evident incapacity to make an active contribution to the community would almost certainly have led to his being abandoned by the rest of his group; he would have been, if not thrown, then at least left to the wolves. However, it is clear that the Old Man of Shanidar was not only allowed to remain within the community but was also nursed back to relative health; all the wounds he sustained while alive show signs of having healed.

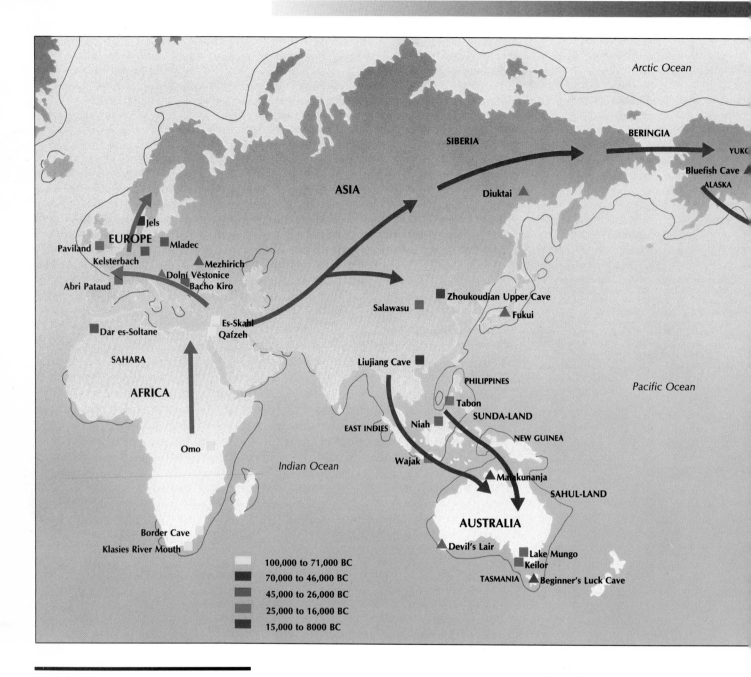

Arctic Ocean

BERINGIA

SIBERIA

YUKO

Diuktai ▲

Bluefish Cave

ALASKA

ASIA

EUROPE

Jels ■

Paviland ■

Mladec ■

Kelsterbach

▲ Mezhirich

Dolní Věstonice ▲

Abri Pataud

Bacho Kiro

Es-Skahl

Qafzeh

Dar es-Soltane ■

Zhoukoudian Upper Cave ■

Salawasu ■

Fukui ▲

SAHARA

AFRICA

Liujiang Cave ■

PHILIPPINES

Pacific Ocean

Tabon ■

SUNDA-LAND

Omo

EAST INDIES

Niah ■

NEW GUINEA

Indian Ocean

Wajak

Malakunanja ▲

SAHUL-LAND

Border Cave ■

Klasies River Mouth

AUSTRALIA

Devil's Lair ▲

Lake Mungo ■

Keilor ■

TASMANIA

Beginner's Luck Cave ▲

100,000 to 71,000 BC
70,000 to 46,000 BC
45,000 to 26,000 BC
25,000 to 16,000 BC
15,000 to 8000 BC

Above are the routes along which it is thought that
Homo sapiens sapiens spread from their African home
between 100,000 and 8000 BC. The earliest sites where
skeletal remains have been found are marked with a
square; the sites of tools and other remains are marked
with a triangle. The blue line shows the world's coast-
lines about 16,000 BC, when a lowering of sea level had
created a land link between Asia and North America.
New Guinea, Australia, and Tasmania formed a continent known to modern geographers as Sahul-land. The
Philippines, Malaysia, and Indonesia were joined with
Asia, forming the continent of Sunda-land.

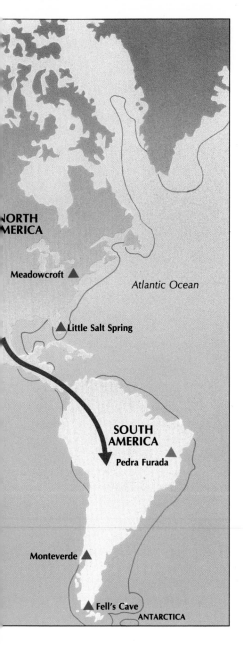

North America

Meadowcroft ▲

Atlantic Ocean

▲ Little Salt Spring

SOUTH AMERICA

Pedra Furada ▲

Monteverde ▲

▲ Fell's Cave

ANTARCTICA

Nor did this caring come to an end when Neanderthals died. Some, at least, were given a respectful burial. Another of the Neanderthal skeletal remains discovered at Shanidar was surrounded by patches of flower pollen, which indicates that this individual was laid to rest upon a mattress of woven pine boughs and flowers, and that, as a final token of esteem, the corpse was strewn with brightly colored blossoms of hollyhocks, grape hyacinths, and cornflowers. It is possible that these flowers, which are employed to this day in Iraqi herbal remedies, were placed with the corpse as a medicinal aid in the afterlife; equally likely is the theory that the blossoms were scattered over the body in much the same spirit that moves present-day mourners to lay flowers by a graveside.

Despite possessing these obvious sensibilities, however, the Neanderthals seem to have had their limitations. For instance, excavations of the places where they lived yield no evidence of any kind of decoration in the form of paintings, engravings, or jewelry. The practical applications of animal bone—for tools—seem to have passed them by, and they also seem to have had little interest in traveling outside their own immediate environment; the stones with which they made their tools came only from the rocks in the immediate vicinity.

Indeed, all the remarkable advances made by these earlier species fade into insignificance beside the explosion of change wrought by *Homo sapiens sapiens*—the anatomically modern humans. No more than 70,000 years after the first appearance of this species in Africa (around 100,000 BC), *Homo sapiens sapiens* were to be found throughout virtually the entire habitable world. They had supplanted all earlier forms of human beings and had colonized totally new areas such as the Americas, which *Homo habilis* and *Homo erectus* had never penetrated. By 35,000 BC, this species had entirely populated the Neanderthals' Eurasian territory.

The fact of *Homo sapiens sapiens'* dominance is unquestioned. What remains unknown, however, is the way in which these early humans managed to achieve that dominance. Did these patently superior creatures, with their glittering accomplishments in toolmaking and hunting, arise on the continent of Africa in 100,000 BC and spread across the world, squeezing out earlier types of hominid, which then became extinct? Or did *Homo sapiens sapiens* simply evolve naturally out of the existing populations of *Homo erectus* in Europe and Asia?

Conflicting answers to this question have been provided by investigations into one particular type of human DNA (deoxyribonucleic acid) known as mitochondrial DNA, found in cell particles that generate energy by burning fats and sugars. Of the two major groups of mitochondrial DNA carried by all human beings, the rarer type is found only in Africa. This much is agreed. However, scientists differ concerning the rate at which DNA has changed over the centuries. Those who estimate one rate of change claim that the branching off toward the non-African type of DNA occurred between 180,000 and 90,000 BC, supporting the theory that a new species evolved in Africa and spread outward, interbreeding with, or even supplanting, earlier Asian and European hominids.

However, this view is challenged by those who put forward a different estimate of DNA's rate of change: that the branching off happened around 1.5 million BC, more than one million years before the emergence of *Homo sapiens sapiens* in Africa. Therefore, according to this theory, the colonizers could not have been *Homo sapiens sapiens,* who did not emerge until around 100,000 BC. Instead, they must have been earlier African hominids, namely *Homo erectus,* who during their expansion evolved first into archaic forms of *Homo sapiens* (including Neanderthals) and then into *Homo sapiens sapiens.*

Whatever the case, *Homo sapiens sapiens* were the only humans who dominated wherever they went. Whole societies were swiftly transformed by the innovations they brought with them.

Their first revolution was technological: *Homo sapiens sapiens* refined blade manufacture with astonishing results. Whereas two pounds of flint had provided *Homo erectus* with a mere four to eighteen inches of knife edge and the Neanderthals with little more than six feet, the *Homo sapiens sapiens* artisans could produce a staggering thirty yards. They achieved this not by hammering the flint core directly with a rock but by interposing a soft chisel point of bone or antler, with which they could coax off large numbers of razor-sharp slices. This process produced sturdy, elongated blades, but *Homo sapiens sapiens* workers could also manufacture blades as small as three-fourths of an inch for precision work. In addition, they made a variety of tools for special uses: scrapers for cleaning hides, and burnishers for smoothing the hides down and rubbing in colorings. These artisans also made saw blades and awls, punches and borers, and they made flint burins, which were tiny points that were used for engraving antler and bone.

Such tools as these, rarities in Neanderthal times, now became commonplace as artisans devised the technique of annealing, which involved making flint more workable by heating it up, perhaps in warm sand. This process dried out the water within the flint, thereby making it more brittle and, hence, more likely to flake in the direction the craftsman's chisel urged. Annealing would have been useful in the manufacture of the new, more deadly flaked-flint spearheads that *Homo sapiens sapiens* devised. Another of *Homo sapiens sapiens'* innovations was the joining of two different materials to produce composite tools, such as stone blades fixed to wooden handles by means of a glue made from tree-bark resin. The addition of handles gave the user a firmer grip and much greater leverage than previously possible with a naked blade. These tools took a long time to manufacture and were designed to provide many years of service.

Homo sapiens sapiens took another technological leap when they began using animal by-products, in the form of bone and ivory, as the raw materials for tools and weapons. Less brittle than stone, both of these materials could be formed into delicate, finely shaped points; sewing needles, for example, could be made out of bone, accompanied by animal sinew for thread. Bone also proved ideal for making barbed harpoons, which were used to spear large fish and water mammals, fishhooks, and fish gorges—straight lengths of bone-splinter sharpened at each end to catch in the throats of biting fish.

With *Homo sapiens sapiens,* the art of weaponry took a stride—a grim stride—forward. Not only did *Homo sapiens sapiens* woodworkers devise the bow and arrow, but they also came up with another device to increase the hunter's killing power. This was the spear-thrower, a short length of wood with a hook at one end that clipped onto the blunt end of the spear. Once the spear was in place, the hunter would balance the spear-thrower on his shoulder and then whip it sharply forward, dispatching the spear toward its target but keeping hold of the spear-thrower by means of a thong handle. Because of the centrifugal force generated by the wrist-snapping action of the hunter, a spear launched in this manner would travel farther than one thrown by hand. Twentieth-century experiments have demonstrated that a hunter using a spear-thrower can hurl a six-foot-long spear up to 150 yards, while a hunter without a spear-thrower is unable to manage more than seventy yards. Nor did *Homo sapiens sapiens* confine the use of these new weapons to killing animals. The skeleton of a child from a cave at Grimaldi, in northern Italy, was found with an arrowhead buried in its backbone; the world's first evidence of human committing violence against fellow human.

It may well have been the development of weapons like this, and of superior tactical ability, that led to the refinement of the *Homo sapiens sapiens'* physiques—they were only two-thirds the weight of their predecessors. Neanderthals had had to be husky. Without the benefit of the spear-thrower, they had been forced to tackle their prey at close range, using their spears more as stabbing tools than as missiles; in this type of confrontation, brute strength would have been essential for survival.

By contrast, the new hunters not only carried new weapons, which allowed them to fight at a distance, but they also devised new strategies, which made it possible for them to kill animals in far greater numbers than could have been achieved by individual, direct combat. Game drives, for example, were now carried out on a scale hitherto unthinkable; whole herds of mammoths, horses, reindeer, ibexes, and other large mammals were systematically driven over the edges of cliffs to their deaths. *Homo sapiens sapiens* even took much of the labor out of searching for game. Instead of roaming across the landscape in the hope of happening on suitable prey, they would simply set up camp at some particularly narrow entrance to a valley and wait patiently for animals to pass by. Similarly, they would set up camp at river crossing points known to be used each year by migrating herds; when the victims arrived, the *Homo sapiens sapiens* would pick them off one by one from dry land as the animals floundered about, terrified, in the water.

Groups of hunter-gatherers now began to organize their way of life more systematically than ever around the migrational patterns of mammals; the seasonal availability of fish or shellfish in the ocean or rivers; and the ripening of particular fruits or vegetables at different times and in different places. Annual timetables evolved to accommodate these key events, and the hunter-gatherers traveled around a fixed circuit of different campsites throughout the year. These groups may still have lived a nomadic existence, but there was nothing at all haphazard about their wanderings.

Characteristically, the members of a hunter-gatherer band would lay claim to a large area of land, different zones of which they would visit in turn. Since there was no way of knowing when resources in one zone might suddenly run low, it was important to keep in routine touch with all parts of the territory, however far flung they were and however productive and comfortable one's immediate sur-

A leaping horse, fashioned from a reindeer antler by a *Homo sapiens sapiens* artisan between 15,000 and 10,000 BC, adds an aesthetic dimension to a utilitarian tool. The spear-thrower, normally a straight wooden rod about two feet long, carved with a groove in which the shaft of the spear could rest and with a hook at one end to hold the butt, acted as an extension of the hunter's arm. When snapped forward over the shoulder, it increased the range and velocity of the spear. This example, only eleven inches long, would not have been practical and was intended as a ceremonial object.

roundings might be. At the same time, even though they were based in different places at different times of the year, the hunter-gatherers developed a strong sense of home. Whereas nonhuman primates had always stayed together when searching for food, these humans probably divided their labor, some of them going off on hunting expeditions that could last for days while others of the group were scouring the area close to the the campsite for nuts and vegetables. Constant scattering of the group made it essential that all of its members have a base to return to, and this need for a secure focal point gave the camps of *Homo sapiens sapiens* a feeling of semipermanence that was lacking in those of their predecessors. While excavations of earlier hominid sites have uncovered signs of domesticity amounting to little more than ash from constantly maintained fires, investigations of known *Homo sapiens sapiens* bases have produced evidence of far greater attention to home activity and comfort.

They constructed their huts out of stone, wood, and the bones and hides of animals. Inside, hearths were placed over small trenches that led out into the open; this ensured a constant supply of air to the flames, the draft acting like a bellows in keeping the fire going.

The search for warmth, in fact, preoccupied *Homo sapiens sapiens* for much of the time in northern latitudes. Born in the tropical regions of Africa, these humans spread into zones colder than their physical evolution prepared them for. It is likely, although no skin or hair survives to corroborate this theory, that in the beginning, *Homo sapiens sapiens* had a dark complexion and relatively hairless bodies—physical characteristics that make life more comfortable in a hot climate but are not as suited to the chill regions of central Europe. Lighter skins were only a gradual, later development. Shivering *Homo sapiens sapiens* found ready-made homes—in the

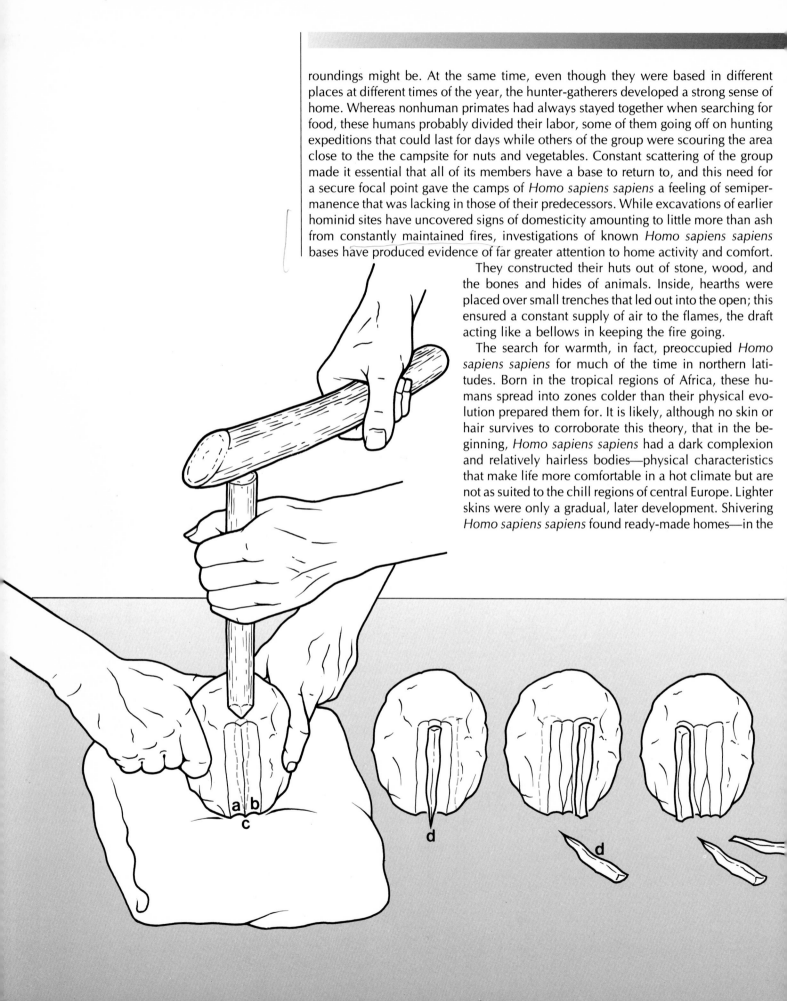

form of natural caves—only in relatively few places, such as southwestern France. Elsewhere, therefore, like their predecessors at Terra Amata, *Homo sapiens sapiens* had to devote a large proportion of their energy to providing themselves with shelter from the elements.

In the long run, this tendency of *Homo sapiens sapiens* toward settling down in one place would result in the evolution of social patterns that are common to all static societies: the beginnings of hierarchy, of rank, of differences between rich and poor. For the moment, though, the values of the hunter-gatherer society were retained in more or less pure form. Living in such small groups, so vulnerable to famine or to attack from predators, so dependent upon one another's labor, hunter-gatherer societies would have tended, at that time as in present-day cultures, to be highly egalitarian. Such bands had no leaders or chiefs; instead, like their modern equivalents—such as the !Kung people of the Kalahari Desert—they employed a system of dispersed leadership, which recognized the particular talents and experience of each participating individual.

Decisions were made collectively, without formality. There was no strong sense of privacy, so grievances were made public quickly and settled without being allowed to fester. The notion of private property, too, was alien to the hunter-gatherer. Instead, the importance of sharing was deeply ingrained: Elaborate rules existed for the distribution of meat and other produce, dependent upon who had killed or gathered it and to whom they were related or obligated. Again, based upon observations of hunter-gatherers today, teamwork must have been absolutely essential: In general, individuals would have disliked standing out among their companions; modesty about one's achievements was a much-prized virtue, and successful hunters would often have gone to extravagant lengths to play down their achievements, however impressive they may have been.

In the hot climates of the modern-day world, where most hunter-gatherers live, the plants and vegetables gathered by the women provide the staple diet; the meat killed by the men constitutes no more than 40 percent of their food intake. For early *Homo*

THE TOOLMAKER'S CRAFT

Simple mass-production methods enabled *Homo sapiens sapiens* to make tools more efficiently. The drawings on the left show how a number of blades were cut from a prepared core of flint: By means of hammering a chisel of bone, antler, or wood, two channels *(a)* and *(b)* were gouged from the flint, thus creating a pointed ridge *(c)*. The chisel was then positioned at the top of the ridge and hammered downward, splitting off a sharp sliver *(d)* and, in the process, creating two more ridges. The process would continue until the whole flint had been used. The slivers were then made into spearheads or knives, or used to fashion more sophisticated products, such as the antler harpoon head *(right)*.

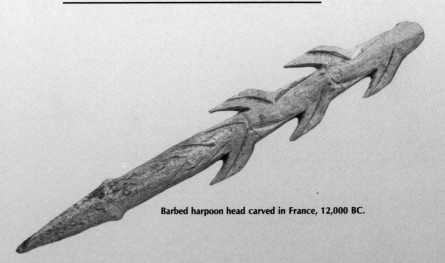

Barbed harpoon head carved in France, 12,000 BC.

sapiens sapiens, this must have been the case, too, although in colder regions, the gathering and storing process would have been far more rushed and labor intensive. In Europe, for example, the annual harvesting season extended over a matter of weeks rather than months; furthermore, staple foods such as beechnuts and hazelnuts had to be painstakingly shelled by hand and provided only small nutritional return in the form of the nutmeat at the center. It is clear, then, that the contribution of the gatherers was of crucial importance. Hunting expeditions depended a great deal upon luck; in the animal world, for instance, wolves manage only one kill for every ten attempts, and even lions fail more often than they succeed. By contrast, the collection of vegetables was predictable in its outcome; one day's foraging by a woman could produce three days' food for her family.

This being the case, it was vital for the survival of the group that women not be kept from their gathering duties by constant cycles of pregnancy and birth. But how to prevent conception? The solution was a simple and probably unconscious form of birth control, based on the principle that breast-feeding mothers have a low incidence of conception—the infant's constant sucking action stimulates the production of prolactin, a hormone that can inhibit ovulation. In contemporary hunter-gatherer societies, women breast-feed their children until they are at least three years old, thereby maintaining a three- to four-year gap between children. Earlier hunter-gatherer societies would probably have evolved a similar pattern; bearing children more often would have made the twin burdens of childcare and foraging impossible to sustain, while bearing children less frequently might have endangered the continuation of the group.

In addition to operating this rudimentary form of family planning, hunter-gatherers must have followed certain patterns of self-regulation for the good of their society as a whole. In order to provide themselves with maximum security and optimum efficiency in finding food, hunter-gatherers today form little bands of between 25 and 30 people, each group including about 12 adults and their children. Approximately 20 such groups are loosely affiliated as a tribe, which, in turn, comprises some 475 to 500 individuals in all. This figure is just large enough to prevent group members from having to look either outside the tribe or among their close relatives for sexual partners, practices that can undermine a tribe's self-sufficiency or create genetic defects through inbreeding. Regular tribal gatherings—held ostensibly for combined hunting or foraging operations—would have provided opportunities for socializing and for meeting possible partners from other groups.

This pattern first established itself during the early days of *Homo sapiens sapiens* as a means of organizing the world's population at a time when it was growing extremely quickly. The welfare of the group was always paramount.

Such an emphasis on cooperation and sharing made communication indispensable for survival. While many mammals and birds have distinctive alarm calls to warn their companions of danger, and while some primates even have several different calls to indicate different sorts of danger, no other species has anything like the linguistic gifts of humans. Studies of the skulls of early hominids have allowed specialists to reconstruct the vocal tracts of australopithecines and *Homo erectus.* They have discovered that australopithecines—like present-day babies—had flat palates, which would have left them unable to vocalize many more sounds than can the apes of today. The palates of *Homo erectus* were a little more curved, allowing them a slightly greater range, although probably not much more than that of a young modern-day child. We do not know for certain whether Neanderthals were capable of truly articulate speech. Most specimens indicate that the Neanderthal larynx, containing the vocal chords, sat much higher in the throat than the larynx of human beings today. This restricted the size of the pharynx—the area above the larynx in which the speaker, by contracting the throat muscles, modulates the sounds produced by the vocal chords. This restriction, therefore, reduced the number of vowels and consonants that the Neanderthals could utter. At the same time, however, Neanderthal remains that display an essentially modern throat structure have been found in the Middle East.

On balance, it seems unlikely that the majority of Neanderthals had any great degree of vocal ability; no examples have ever been discovered of more refined Neanderthal endeavors in which speech and the communication of ideas would have played a part.

It was only with *Homo sapiens sapiens* that the palate became fully curved and that language began to attain any richness. Two new developments made this possible. First was a gradual increase in human brain capacity and, with it, that part of the brain that translates thought into speech. And second was the growing number of cultural and social demands imposed by the increasingly complex nature of the communal society in which *Homo sapiens sapiens* lived. When humans first walked the planet, individual actions, however ingenious, had not demanded the extensive use of language. None of early *Homo sapiens sapiens'* toolmaking breakthroughs, for instance, had depended on the capacity to speak. Neither had developments in hunting technique: Modern-day hunter-gatherers still tend to do their hunting in silence, and in any case, other animals—African wild dogs, for example—can perform highly elaborate group maneuvers without the benefit of language. Linguistic competence did not become necessary until humans started to live in settled, organized groups, where they evolved systems of planning, of allocating tasks, of sharing meat and other produce—systems that went beyond the immediate, self-evident needs of the moment to cater to what was not yet apparent, not directly obvious: for what, in short, needed to be explained.

As the linguistic ability of human beings developed, it took them further and further into the realm of the abstract and the symbolic, opening up a world beyond the here and now. Previous artistic activity seems to have been limited to the adornment of practical objects. Suddenly, however, around 30,000 BC, there came a great outpouring of creativity across the populated world. An astonishing range of portable artworks—strange clay and stone figurines in human and animal forms, bone and ivory pieces engraved with dazzling virtuosity, delicate beads, shell necklaces, and other ornaments—and more permanent items, such as the awesome cave paintings of France and Spain, provide eloquent testimony to this new, artistic sensibility. The purpose of such works remains a mystery: No one knows whether the motivation behind them was religious, educational, or aesthetic.

One of the few conclusions that can be drawn from the cave paintings is that most early human beings were right-handed. Observation of hand stencils painted on the walls shows that the majority of hands depicted are left hands, which suggests that the artists used their right hands to apply the paint; at the French cave of Gargas, for example, there are 136 left hands on the wall, compared with 22 right hands. But even though the cave paintings pose more questions than they answer, there can be no doubt that they and other examples of early art indicate a human race more than ever before in control of the immediate necessities of physical existence and, therefore, possessed of the leisure and confidence to develop a richer cultural life.

Homo sapiens sapiens' advanced system of social organization helped equip them for the enormous range of environments they encountered as they expanded gradually outward from their African origin, up through the Sahara and North Africa into central Europe, and then right across Asia to Siberia and the Americas, in the north, and to Indonesia and Australia in the south. The impetus behind this expansion was the constant search for better food supplies—a search that continued until all the regions of the Old World and, ultimately, the regions of the New World had been populated by humans.

That they succeeded in colonizing the globe was due to their capacity for teamwork. Their highly developed, deeply ingrained systems of cooperation and interdependence made them far more flexible than earlier hominids and, thus, able to make the most of whatever situation they found themselves in. A look at the way some of these earliest modern humans lived, in sharply contrasting settings, illustrates this remarkable adaptability.

On the coast of southern Africa, in a cluster of caves and crude rock shelters at the mouth of the Klasies River, lived some of the very first communities of *Homo sapiens sapiens,* long before the species appeared in Eurasia. Between 100,000 and 70,000 BC, as sea levels fluctuated through glacial cycles, the site remained close to the coast; after that, the waters receded, ultimately to a distance of about forty miles, and the complex was abandoned.

However, for those 30,000 years, successive generations of these settlers led a favored existence. Throughout this period, the region enjoyed mild climatic conditions and abundant food supplies from both the land and the sea. Groups of women foraged in the bush, while smaller groups of men—two or three at a time—made longer journeys inland, hunting the big mammals on which the entire community could feast: blue antelopes, Cape buffalo, black wildebeest, and the now-extinct giant buffalo, which might have weighed as much as 3,500 pounds. Some species seem to have been avoided—no remains have been found of the choleric bush pig, for example, which was always liable to turn ferociously upon its attacker. It was probably regarded as being too dangerous to be worth hunting.

In addition to satisfying immediate food needs, the meat from these kills would also have fed the Klasies River community during the winter, when meat was scarce; any leftovers would have been dried in the smoke from their fires, a simple but effective way of preserving meat in a warm climate. The cave dwellers also used fire to improve their yield of vegetable crops; this technique involved burning patches of the bush to stimulate the growth of smaller food plants, which otherwise might have been choked by larger, inedible vegetation.

The people of the Klasies River mouth did not look only inland for their provisions. The myriad rock pools along the shore were systematically combed at each low tide for limpets, periwinkles, and mussels. Seabirds were killed for food, and there is evidence that sea mammals—whales and small dolphins—were also eaten, probably after being washed ashore. Cape fur seals, wintering on nearby beaches, were often slaughtered, although probably as much for their fur as for their meat. The ocean, therefore, proved an important source of nourishment, in particular because it sup-

These drawings reconstruct the type of dwelling that was built by *Homo sapiens sapiens* hunter-gatherers on the treeless steppes of the Ukraine about 13,000 BC. Approximately sixteen feet in diameter, the framework was built entirely of mammoth bones and weighed some twenty-three tons *(left)*; arching tusks formed the doorway, and each component was held in place by pins made from smaller bones. A covering of turf, reinforced by loosely sewn furs and skins *(below)*, with further layers added according to weather conditions, turned this domelike structure into a cozy home through the long, harsh winter months.

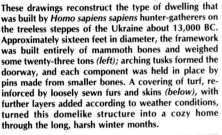

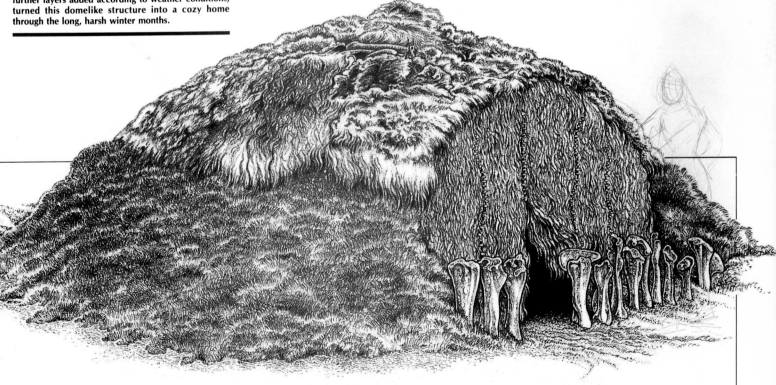

plied much-needed sustenance during the winter months, when edible plants were more difficult to find. The addition of seafood to the basic hunter-gatherer diet transformed this community's position at a stroke, giving them an extra layer of security against food shortages.

Although the settlement at the mouth of the Klasies River was abandoned around 70,000 BC, the people who lived just down the coast at Nelson Bay—some 60,000 years later—had clearly absorbed many of the lessons learned by their forebears. They, too, lived in and around a broad-mouthed cave, which had fresh water conveniently trickling from a spring at the back. Like the Klasies River settlers, they also went hunting inland, although they now felt competent to tackle bush pigs and warthogs and had added ostriches and baboons to their diet. Technologically, though, they had advanced well beyond their predecessors: They made more sophisticated tools; they built stone rims for their hearths; and they erected a windbreak constructed of wooden posts draped with hides or brushwood to protect the entrance to their cave.

The people of Nelson Bay had developed a number of efficient methods of fishing, the most notable of which was the fish gorge. This small, sharpened sliver of bone or wood, perhaps two inches long, was tied to a line made of either leather or animal sinew. The device lodged sideways in a fish's throat when swallowed, allowing the fisherman to haul in his catch. Later, the gorge was superseded by the fishing net, also thought to have originated in South Africa. This device was woven from vegetable fiber or hide thong and was weighted with specially prepared cylindrical stones; the arrival of the net marked a crucial breakthrough in that it made large-scale harvesting of fish possible for the first time. Thus, the sea became a reliable and regular source of food, enabling coastal communities to stay in one place for most of the year.

In the meantime, 3,700 miles to the north, several separate communities had, from around 15,000 BC, been developing a very different way of life. They lived on the Kôm Ombo plain, a 155-square-mile stretch of land on the eastern bank of the Nile River, in what today is southern Egypt. An area of boggy pasture for much of the year, the plain offered lush reeds and edible grasses that were much to the liking of the large herds of wild cattle that inhabited it. During the hot, dry months between March and August, before the annual rains began, the cattle were joined by antelopes and gazelles, which came to the plain looking for food and water.

At first, it was the antelopes and gazelles that attracted hunter-gatherers to Kôm Ombo; once there, however, the hunters discovered that the plain teemed with other creatures, especially during the period between August and October, when the rains in East Africa built up a large head of water upcountry and sent wide river channels rushing across the plain. In addition to the wallowing hippopotamus, a formidable source of meat when it could be overcome, the hunter-gatherers found a rich variety of edible water birds, both resident species and migrants passing through. Among the types of birds the hunter-gatherers could choose from were ducks, geese, cormorants, and even herons and eagles. Meanwhile, beneath the surface of the water, schools of catfish, perch, and other freshwater fish abounded, along with the soft-shelled turtles of the Nile.

Such natural riches allowed several hundred people to occupy this relatively small area at the same time—a remarkable population density, given the hundreds of square miles needed to support a similar group in less accommodating environments. Following traditional hunter-gatherer patterns, the people of Kôm Ombo lived in groups of between forty and fifty people, each group retaining its own specialized way of life, identified by its own slightly different tools and manufacturing techniques. Some bands had become so settled that they even practiced a very basic form of agriculture, gathering grain on an increasingly systematic basis from the cereal grasses that grew all over the plain.

Altogether more hostile conditions prevailed on the steppes of central Europe and Asia, where long, bitter winters brought temperatures of -35°F., and short-lived summers never got warmer than 70°F. Even here, however, *Homo sapiens sapiens* survived and even thrived.

The steppe held obvious appeal for any hunter-gatherer band: herds of reindeer, bison, horses, mammoths, and woolly rhinoceroses. But rich as the steppe was in wildlife, it presented a decidedly unfriendly environment in other respects. Trees and plants grew only sparsely, and without significant hills or forest cover to provide natural screens, the biting winds tore across the plains unhindered. Shelter was of paramount importance in such conditions, but there were few caves to provide natural protection from the elements; so human beings had no choice but to construct their own housing, using whatever materials were at hand. In parts of eastern Europe where wood was scarce, mammoth bones were a common substitute.

Indeed, for the community that around 25,000 BC settled semipermanently at Dolní Věstonice, in southern Czechoslovakia, life seems to have revolved almost entirely around the mammoth. From the large, artificially shaped hollows that now indicate where the settlement's four huts stood, come finds that demonstrate the beast's importance to the members of this community. It seems that they used mammoth hides to cover the huts and mammoth bones to hold down those hides. Mammoth tusks and bones provided the framework for a defensive stockade around the huts, and the settlers overcame the shortage of fuel wood by burning mammoth bone instead. Although difficult to light initially, the bone gave off a hot flame once it was burning and must have provided a cheering warmth inside the largest of the four huts, a 990-square-foot building in which as many as five hearth fires might have been lighted at one time. It was around these hearths that the people fashioned small models of the lionesses, rhinoceroses, and other animals they encountered out on the steppe; some were made of clay, others—inevitably—of ivory mammoth tusk.

Originally, the Dolní Věstonice site had been established as a base for hunting the mammoth. The site chosen by these hunter-gatherers for their village must have lain very close to one of the chief routes used by the mammoth herds on their winter migration southward. Each year, as the great beasts moved down the valleys of the Dyje, Morava, and Danube rivers, the hunters lay in wait for them. There was little that was noble or stirring about the mammoth hunt—no epic struggle for mastery between man and beast. Instead, every autumn, year in and year out, the hunters ambushed the herds at the same spot, raining arrows and spears upon the terrified animals and then butchering them on a near-industrial scale. While the hunters took as much meat as they could from their kills, they inevitably left large amounts behind, because the massive carcasses were too heavy to move. The meat that the hunters did manage to cut free was either dried in the wind and sun on a nearby hillside or else stored in frozen pits.

In addition to the killing of mammoths, the people of Dolní Věstonice got their meat by scavenging—that oldest of hominid strategies—and by hunting down the other large mammals that grazed on the steppe, including reindeer, horses, and bison. From the carcasses of these animals, they derived not only meat to eat but also hide with which to cover both huts and human beings.

The clothing of the Eurasian hunter-gatherers most probably resembled that of the Inuit today: hooded tunics made of hide or fur, with tightly sewn seams to trap the heat, and leggings that could be tucked snugly into hide boots. Fur socks and mittens gave the feet and hands further protection. The fur came from hares, foxes, and martens, which the hunter-gatherers caught in rudimentary snares: rawhide nooses held taut by bent saplings, which sprang up when disturbed by a passing animal, jerking their victims into the air and leaving them dangling there until frostbite or returning hunters applied the *coup de grâce*. Birds, too, were hunted, either caught in traps on the ground, or brought down in midflight with arrows or other missiles.

In the spring and summer, the people of Dolní Věstonice had to compete for their prey on equal terms with other carnivorous animals. However, in the winter, the humans' mastery of fire gave them a clear edge over their rivals, since it enabled them to thaw out the carcasses of animals that had died in the snow, rather than having to wait until nature did the job for them in the spring.

Not that they would have starved if no meat was forthcoming. Like other communities of this region at the time, the people of Dolní Věstonice probably supplemented their meat diet with fish from surrounding rivers: Pike and perch were both popular, but salmon quickly became the staple fish, once its habits were understood. Similar to the mammoth, it was rendered vulnerable by the predictability of its annual migration. Each year, salmon swimming upstream to spawn had to run the gantlet of predatory humans, who flocked to the riverbanks at the appointed time to catch the migrants. Often the hunters blocked the rivers with weirs built of stones, channeling the fish into confined spaces where they could be speared easily, perhaps with the leister, a Viking word meaning "stabbing tool." This weapon was an ingenious three-pointed spear, the central tine of which impaled the fish, while the two curved prongs on either side ensured that it did not wriggle free.

The salmon-hunting season lasted only a short time. It took fewer than three weeks for all the fish to pass by a particular point, and some 40 percent of them would swim past within the space of just three days. It was vital to make the best of this critical time, and so, while the men fished feverishly against the clock, the women would frantically gut, dry, and store the catch. Theoretically, the salmon run should have been a major source of food; however, the sheer scale of the operation usually outweighed the physical resources of a small, hunter-gatherer band, whose members would be unable to grab hold of more than a small proportion of these showered riches. This limitation meant that salmon could never be more than a supplement to their main diet of meat.

For the inhabitants of Dolní Věstonice, food remained, of necessity, the first priority. However, they did not neglect the finer things in life. There was certainly music, either as an accompaniment to religious ritual or as entertainment; someone in the community played the simple flute, made from hollowed-out bird bone, that was found at the site many thousands of years later. This musician was perhaps accompanied by others, playing some of the mammoth-bone percussion instruments of the time: the xylophone made of hipbones, drums made from shoulder blades or hollow skulls, and rattles made of jawbones. The visual arts were also provided for, although the great, blank cave walls of France and Spain were not available to the Czecho-

MARKS WITH MEANING

The possibility that humans may have devised a system of recording information as early as 30,000 BC is suggested by the marks on a section of reindeer bone recovered from a site in the Dordogne in France. The whole bone plaque is shown actual size at bottom; the enlargement immediately below is overlaid with the outlines of the sixty-nine engraved marks. After making an analysis of these impressions, archaeologist Alexander Marshack concluded that, far from being random doodles, they were purposefully created, possibly with a number of different tools and over an extended period of time. Relating the marks to the waxing and waning of the moon, Marshack proposed that they constituted a consecutive record of lunar phases.

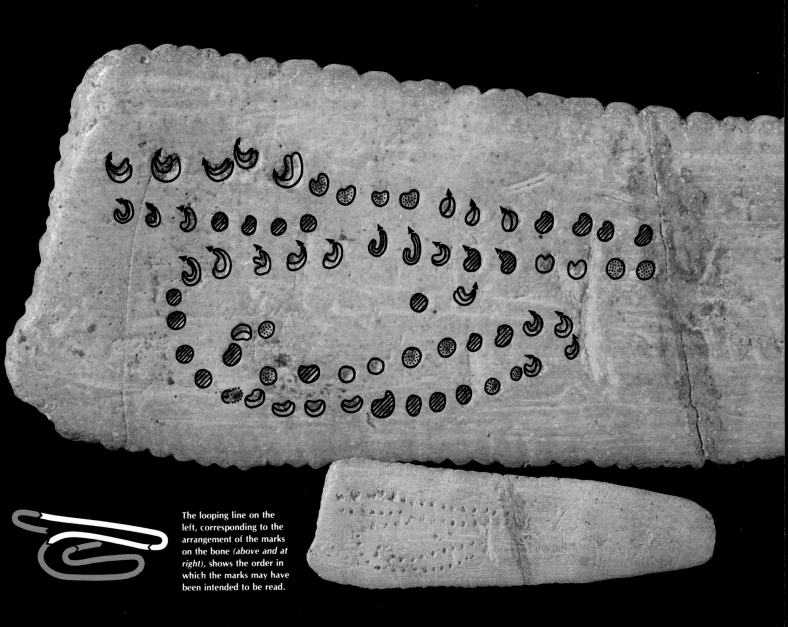

The looping line on the left, corresponding to the arrangement of the marks on the bone (above and at right), shows the order in which the marks may have been intended to be read.

slovak artists of the time. Instead, works of art were formed in terra cotta: A fifth hut at Dolní Věstonice, standing apart, outside the stockade, housed the world's first known pottery kiln—an oven enclosed within earthen walls in which sculptures of soft clay, mixed with powdered bone, could be fired into more lasting forms. These first potters made animal heads—foxes, bears, and lions—and they fashioned strange statuettes of female figures with exaggerated, swelling breasts and buttocks. Fascinating as these objects are aesthetically, they carry still greater significance in terms of the technical progress they represent: The potters' clay is the earliest example of two different substances being compounded to produce a useful third material.

Inevitably, the success of *Homo sapiens sapiens* in adapting to such a wide range of environments led to a boom in population throughout the closing millennia of the Ice Age. As an area threatened to become crowded and its resources overexploited, whole hunter-gatherer bands—or smaller splinter groups, perhaps of young people seeking greater independence—would move out in search of virgin territory. Little by little, then, the earliest modern humans spread across the whole of the Old World, acquiring new skills and customs as they went.

It was in this way that Australia was first colonized, around 50,000 BC, by peoples who had gradually moved southward during many thousands of years, progressing through what is now the island region of the East Indies but what was then a continuous expanse of dry land. We do not know whether the route these people took led them through southern China and the Philippines, or from Southeast Asia to Indonesia. We do know that the next stage in their journey must have been a sea voyage of almost sixty miles across the Java Trench, the deep expanse of water that stood between them and the neighboring continent of Sahul-land, which, up until 8000 BC, joined Australia, New Guinea, and Tasmania in one landmass.

What sort of vessels these early mariners used remains a mystery. Certainly, more recent Australian Aborigines have had no boats equal to the task, which leads to the theory that bamboo—native to Southeast Asia but not to Australia—was used in their construction. But whatever the composition of their boats, and however flimsy or makeshift they might have been, it seems clear that the purpose of these first Australians was both clear and well formed. They were seeking a new life in a new land, and they were prepared to set out across what might have been, for all they knew, an endless tract of ocean in order to find it. Mathematics reinforces this theory: Had one lone raft arrived accidentally, as a result of navigational error or being blown off course, its passengers would have been unable to parent children in sufficient numbers for the later population to reach the size it did. It seems certain, then, that deliberate migrations of several rafts took place, probably in several phases.

Initially, the settlers remained on the coast. The sandy beaches and mangrove swamps of northern Australia were very similar to those of Indonesia's low-lying islands; and the crocodiles, turtles, seabirds, fish, and shellfish, as well as the fruit trees and other vegetation, also closely resembled those that their ancestors must have hunted and gathered. The new settlers maintained this lifestyle as they worked their way around the coastline: It was the way of life they knew best, and in any case, the Australian interior seemed remarkably unenticing, with its giant marsupials and strange flightless birds.

However, having reached the southern coast of the continent, some bands began working their way inland—following the great Murray-Darling river system—where they established a freshwater version of their old coastal life around the shores of Lake Mungo. Here, for nearly 20,000 years, the lake dwellers lived a comfortable existence, gathering freshwater mussels and catching the carp, cod, and crayfish that abounded in the lake. Brief sorties into the interior brought back emu eggs, wombats, wallabies, and rat kangaroos—small, rat-size hopping marsupials. But around 16,000 BC, the climate in Australia grew increasingly dry, culminating in a drought that dried up Lake Mungo. Like other Aborigine bands across the continent, the erstwhile lake dwellers now had to adapt to a desert climate. Stricken with perpetual drought, vast tracts of land stretched out featureless and bare, save for dried, twisted trees and occasional patches of parched scrub.

Faced with this forbidding environment, the inland Aborigines soon established a way of life that was centered on the hunting of kangaroos and other marsupials, along with the harvesting of wild millet seeds. Not long after this, the giant marsupials became extinct, probably because bush fires, which were deliberately started by the Aborigines and were intended to stimulate the growth of their own food plants, destroyed the shrubs on which the animals fed. When, in 8000 BC, the end of the Ice Age resulted in a flooding of the more hospitable coastal settlements, forcing the inhabitants back into the arid zones, it marked the beginning of an inland, desert-based way of life that would last for thousands of years: Today, the memory of any earlier coastal existence has long since faded among the Aborigines of the interior.

Faintly discernible amid a riot of lines engraved on a stone plaque, a horse's head strains against a crude halter *(left)*; the simplified outline is shown in the diagram below. Excavated at La Marche in France, this stone image suggests that horses were being ridden or led in harness as early as 12,000 BC. Other archaeological evidence suggests that ancestors of the domesticated dog may have been trained to stand guard over the settlements of hunter-gatherers.

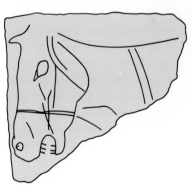

Compared with this epic story of a people's setting out across the ocean in search of a new continent, the arrival of *Homo sapiens sapiens* in North America seems to have been a low-key event. There are two main theories of how human beings first arrived in the Americas. According to the traditional version, hunter-gatherer bands who had settled on Beringia, the icy stretch of land that, from 25,000 BC, joined Asia and America, simply moved east in search of fresh resources. Had they not already made this move, they would have been forced to make it toward the end of the glacial period, around 13,000 BC, when the sea level rose, and once again, the waters closed over the top of Beringia.

Once in North America, these new immigrants found little to surprise them: Animals, too, had crossed the land bridge from Siberia, and plant seeds had also been blown over, both recently and in the more distant past, when previous glaciation cycles had thrown the route open. For these first settlers, therefore, life went on much as before, except that, to the south, there now stretched the whole of the Americas, a land of almost limitless opportunities for hunter-gatherer peoples.

For a long time, however, the path leading to those opportunities was blocked; beyond Alaska and the Yukon Territory, two vast ice sheets almost met to erect a well-nigh impregnable barrier. One, known today as the Laurentide, inched southward and westward across northern Canada and America; gouging and scraping its slow, laborious way across the landscape, it had joined its counterpart, the Cordilleran, whose frozen heart lay in the center of the northern Rockies. Although some brave souls must have made the hazardous journey through a narrow, ice-free corridor that separated the two sheets, people would not have migrated southward in any numbers until the ice began to retreat, around 10,000 BC.

Once the way was clear, though, *Homo sapiens sapiens* flooded south, spurred by the astonishing richness of game the virgin continent provided. As usual, the hunter-gatherer bands kept moving onward in search of new resources, constantly occupying new territories. They were rarely disappointed: These were times of extraordinary plenty. The mammoth, the bison, and the other large grazing animals that teemed in the heartlands of North America, unaccustomed to the ways of humans, were far more easily hunted than their relatives on the steppes. Game drives proved to be devastatingly effective in these circumstances, and bison and other large mammals were stampeded over cliffs by the thousands, with much of the meat that resulted simply being left to rot. In the course of this bonanza, several species were quickly hunted to extinction, including the wooly mammoth, the ground sloth, and the elephant-like mastodon.

Meanwhile, the human population, its numbers boosted by such prosperity, continued the slaughter on an even larger scale. Resources in each new territory started to run low rather quickly, and the search for new hunting grounds soon took the first humans through to the Central American isthmus and on to South America. By 10,000 BC, the first hunter-gatherer bands had reached the southern tip of Chile, and the occupation of the Americas from top to bottom was complete.

This traditional theory of American settlement insists upon a surprisingly late date of arrival and a remarkably swift rate of expansion through the Americas after that date—some 12,000 miles in 1,500 years. Supporters of the most popular rival theory claim that this traditional version of events sits uneasily with the good deal of evidence of human activity in South America that seems to suggest *Homo sapiens*

sapiens arrived on the continent many thousands of years earlier. Previously, a precise date had proved difficult to establish, since no sign of human life before 12,000 BC has ever been found in North America, through which the South American settlers must surely have passed. Recently, however, much firmer evidence of early settlement in South America has been discovered; charcoal from a hearth found at Pedra Furada, in eastern Brazil, has been shown to be about 40,000 years old.

Advocates of this alternative view believe that the first people arrived in America not during Beringia's final appearance above the waves, from around 25,000 to 13,000 BC, but during its previous appearance, between 50,000 to 40,000 BC. Having been deterred from heading inland by the great mountain ranges of the West Coast, the settlers would naturally have stayed close to the coastline, having no real chance of venturing into the interior until they reached the lower areas of the Central American isthmus. The same rise in sea level that drowned Beringia in 40,000 BC would also have covered the coastal strip of North America down which they had traveled, along with all traces of their occupation.

Whichever of these two theories is correct, it is clear that no one race of *Homo sapiens sapiens* was responsible for the colonization of the Americas. By 10,000 BC, distinct racial groups were already evolving. Although dental evidence from skeletal remains confirms the view that native Americans all originated in northeast Asia, in the Siberian triangle formed by Beringia, northern China, and Lake Baikal, it also corroborates the theory that the original colonization consisted of three different waves. First, there was a large initial migration, producing all of the native South American peoples and most of the later North Americans. Then, several thousand years later, came a second, smaller wave, made up of the ancestors of the Apaches and the Navajos, as well as the indians of the Alaskan interior and parts of British Columbia. Finally, a third group of coast-dwelling peoples, the earliest Inuit and Aleuts, arrived to add their genes to the fast-expanding American pool. The basis for this three-wave theory is the similarity of the languages spoken by the different peoples in each of these groups; the linguistic evidence suggests that although North American Indians are separated by an entire continent from South American Indians, their languages are similar enough to indicate a common ancestry and, therefore, a common arrival date in America.

But these groups themselves began to diverge once they had settled down and embarked upon different lifestyles. From the pine forests of the Pacific Northwest to the swamps of Florida, from the Great Plains to the Amazon jungle, from the woods of New England to the high mountain ranges of Peru, a wide diversity of habitats demanded widely differing survival skills. Isolated from one another, the inhabitants of such far-flung locations began to develop very different cultures and, moreover, to develop a sense of what distinguished them from people of "foreign" cultures. The idea of ethnicity was taking root. It had begun gathering momentum in the very early days of *Homo sapiens sapiens* in Africa and Eurasia, but as time went on, it would grow stronger and stronger.

It was not merely a matter of cultural difference: Important physical variations were already marking the different races. The massive population increase, as humankind thrived across the world, was in itself a cause of change, each extra breeding person increasing the opportunity for genetic mutations and new ideas to occur. Any small group splitting off from a larger one would inevitably miss some components of the

BOLD IMAGES OF WOMEN

Exaggerated in their proportions, yet conforming accurately to the distinguishing features of the female anatomy, examples of statuettes found scattered throughout much of Eurasia testify to the feelings of awe and mystery experienced by their makers, who lived between 35,000 and 8000 BC. Most of the figurines are made of stone, baked clay, or ivory, and measure from one and one-third to eight inches high. Known as Venus figurines, they represent women at every stage of life from puberty to old age, although the majority—including the ones shown here—are of mature, well-nourished females whose bodies show the effects of age and childbearing. Archaeologists have speculated that the figures may have represented either deities in fertility cults or symbols of authority in matriarchal societies. Although they clearly had a vital role in the religious or social life of the hunter-gatherer groups that created them, their exact function remains obscure.

The protruding stomach and buttocks of this Italian stone figure are visible only from the side.

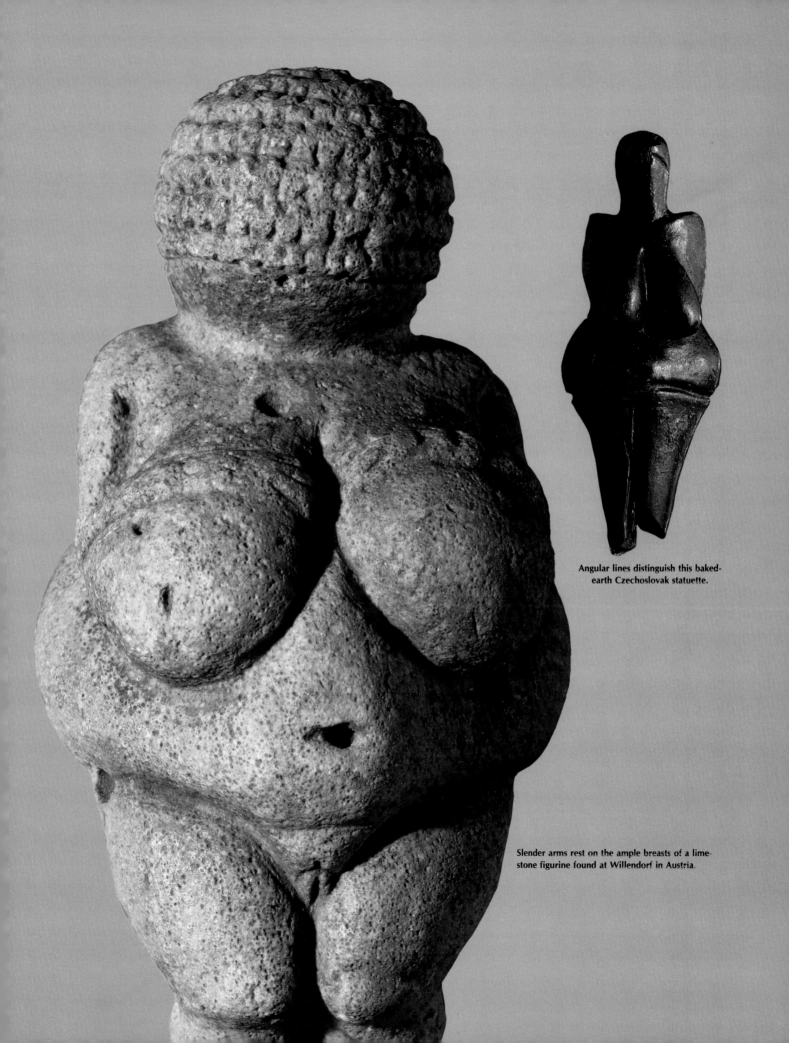

Angular lines distinguish this baked-earth Czechoslovak statuette.

Slender arms rest on the ample breasts of a limestone figurine found at Willendorf in Austria.

original group's genetic pool: Over several generations, breeding within the new group reinforced and exaggerated these differences, with the result that the new group grew ever more distant from its ancestors. As *Homo sapiens sapiens* spread out across the world, coming to grips with a wide range of different environments, this evolution was reinforced by a process of natural selection, which ensured that those genetic variations that suited a group's circumstances were retained, while those that did not were discarded. A tall, slender physique equipped the African for life in the sunbaked bush, since it presented a large area of skin to be cooled by the air. However, the same body shape would have led to a prodigal squandering of heat in an Arctic climate; the Inuit, therefore, developed a much shorter, more compact form and grew thick, straight hair, physical characteristics that helped retain heat.

After a period of less than 100,000 years of their first appearance in southern Africa, *Homo sapiens sapiens* had conquered almost the entire globe. It was, in many ways, a tyrannical rule. From defenseless victims, human beings had evolved into wanton agents of destruction. Whereas they had at one time scavenged the odd carcass or, at best, managed to kill the odd beast, they now drove vast herds to their deaths, with a minimal amount of exertion. Such a reign of terror could not last forever, and as many species were hunted to extinction and others became far rarer, major changes in strategy were required. Even during the Ice Age, there had been signs that humans were beginning to husband their resources. Hunter-gatherers who followed reindeer and ibex herds on the continent of Europe were beginning to exercise at least some loose control over them, corralling selected beasts and domesticating some of them for their hauling power.

Thus, when the Ice Age gave way to a new, warmer era, around 8000 BC, the first stirrings of an agricultural revolution were already starting to make themselves felt at several different points around the globe. After two million years, throughout which they had instinctively followed the hunter-gatherer lifestyle of their forebears, people all over the world had reached the stage in their evolution where they would start abandoning the spear for the plowshare.

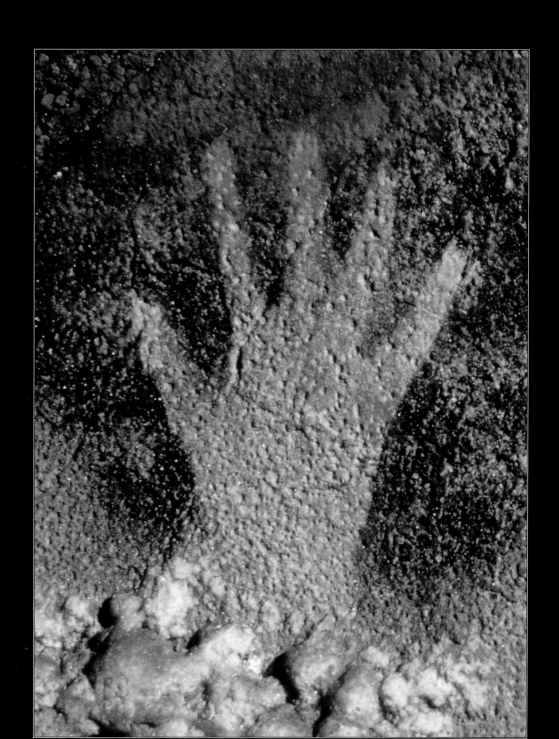

Nowhere is the life of the hunter-gatherer more vividly recorded than in the abundant cave paintings and engravings of France and Spain. Created between 35,000 and 8000 BC, as the last Ice Age peaked and then subsided, this rich legacy of images provides a graphic testimony to the sensibilities of early human beings and to the extent of their cultural development.

Centuries of darkness have preserved the freshness and vigor of these works. But brief glimpses of how *Homo sapiens sapiens* viewed the world about them raise as many questions as they answer. Were the images on the wall thought to contain a mystical or magical power? Are they symbols of some early religion or supernatural belief? Or were they simply educational aids—a kind of pictorial encyclopedia?

Three main subjects are portrayed, the least common being humans, who sometimes appear as stick drawings and sometimes as fanciful combinations of bird or animal and man. The clearest human images are hand prints; the example shown on the preceding page, from the Pech-Merle Cave in the Lot in southwestern France, perhaps served as a signature or a mark of territorial ownership.

Abstract symbols show up frequently, too. These take the form of lattices, boxes, barbs, and rows of dots that may be an early form of hieroglyphics or an attempt to portray inanimate objects such as gates, huts, weapons, or geographical landmarks.

The most frequently depicted images are those of animals. These are mainly confined to four species—three of which appear in the painting below, from the Hall of Bulls at the Lascaux Cave in France's Dordogne, with the addition *(far left of picture)* of a mysterious unicorn-like creature. Here, as in most cave paintings, the animals are shown in profile only.

No one knows for certain what the purpose of these images was. One clue to the mystery, however, lies in the inaccessibility of the paintings, which are often sited in alcoves and recesses reached by means of tortuous networks of underground tunnels. The caves that contained paintings frequently were uninhabited, which suggests that the intention of the artists was not simply to provide a public art gallery. Also, many of the works are located in positions within the cave that would have presented great difficulty and discomfort to the artist. Far from being located at eye level on a handy, upright wall, many paintings are situated several feet off the ground and would have required their creators to have worked either while lying on their backs— perhaps on a rudimentary form of scaffolding—or while clinging precariously to a steep rockface.

At Lascaux, the artist or artists used the entire chamber for their composition, incorporating the natural curves of the walls and ceiling to bring life to the animal im-

ages. The physical obstacles involved in adorning this unusual canvas suggest a motivation deeper than the simple desire for artistic self-expression or display.

One possible interpretation is that the paintings functioned as sympathetic magic. Most of the animals portrayed appear to be healthy, well-fed specimens; a few even show signs of being pregnant, suggesting that the act of painting may have been intended as a charm, designed to promote health and fertility in the animal population—and thereby, among the humans who required the animals for food. In the same way, by depicting beasts being killed with spears and arrows, the artists might have believed they were improving their chances of successful hunting.

The role of the artist may have extended into the realm of the supernatural in another way. It is possible that painters were mediating between the hunters and the other world, conjuring up "spirit animals" on the wall in order to heal a wounded hunter or to help in the tracking of prey.

Such theories are reinforced by the suitability of these dark labyrinths to the staging of religious or ritualistic drama. Some experts believe that the animal images bore a significance to early human beings related not to species but to gender: One theory posits that the bison or aurochs represented femaleness and the horse maleness, their relative positions in the cave spelling out symbolic messages, perhaps having to do with birth and reproduction.

Less ambitious is the idea that the artists used caves in the same way that teachers use blackboards. Thus, the contours of cave walls and ceilings might be used by the artist to portray the surrounding landscape; also, seemingly enigmatic rows of dots may have been intended to mark the boundaries of a hunting territory. Paintings may simply have served as a means of informing fledgling hunters how to identify various types of prey; in some instances, specific features of an individual animal have been highlighted by a little twisting of the perspective, so as to display more clearly the detailed formation of an aurochs' hoof or a deer's antlers. To spellbound children, crouching in a confined space and peering through the flickering flames of a stone lamp fueled by sputtering animal fat or a hearth fire, such pictures would have stayed in the mind for many years.

The only sure thing seems to be that, although they are indecipherable to us, the messages conveyed by the cave paintings were understood both by the people who painted them and by those for whom they were painted. Using the animal figures that appear on the cave walls, the following pages illustrate some of the evidence on which the different theories concerning cave art are based. But these constitute merely the pieces of the puzzle. How they fit together remains a mystery.

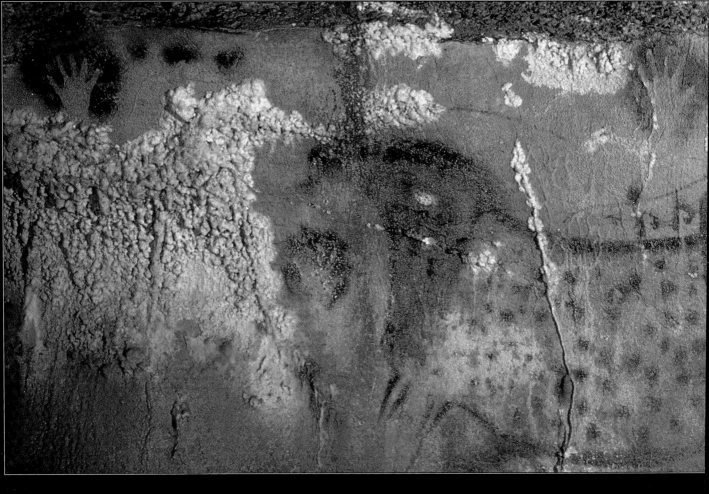

Two graceful galloping horses, both from caves in southern France, bear a distinctive M-shape on their sides, a device used to portray the change of color between the dark hide and lighter belly area. Horses were the most commonly depicted animals in cave art, and these accurately executed paintings demonstrate the artists' familiarity with their subject. The creator of the "Chinese" horse on the facing page, so named because of its resemblance to horses in Oriental art, has invested the animal with a sense of movement and three-dimensionality by leaving the legs on the far side slightly detached from the body. The so-called circus horse on this page, from Le Portel Cave, takes its name from its forelegs' lively high-step.

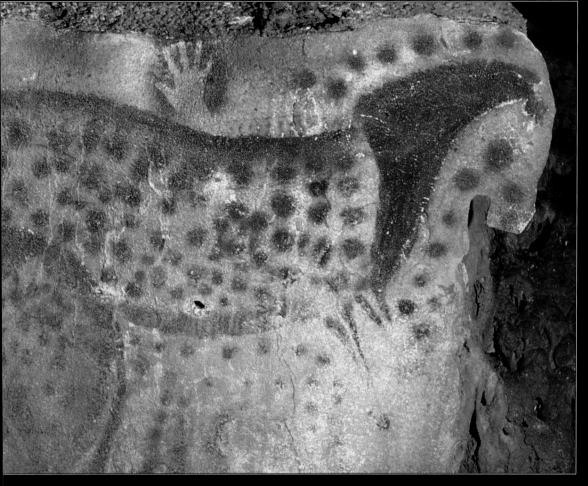

Peppered with dots, two horses stand back to back in a painting from the Pech-Merle Cave. Drawing on the ingenuity that these awkward work surfaces required, the artist has used the natural shape of the rockface to create a sculpture of the right-hand horse's head, while at the same time painting the same horse's head in much smaller, more stylized form, as mirrored by the head of the left-hand horse. The dots and hand prints were probably made by the artists' filling their mouths with liquid paint and then spraying or spitting it onto the walls.

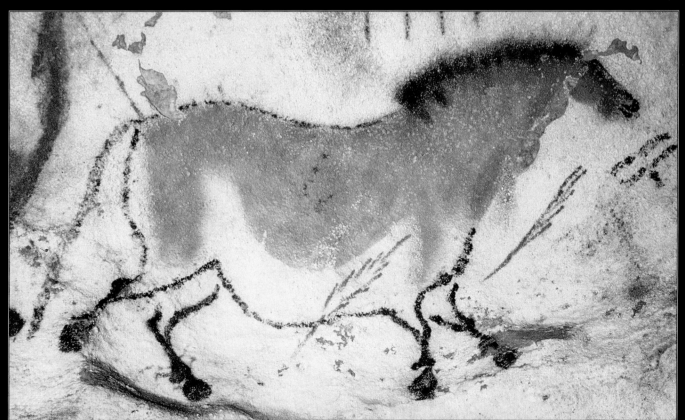

With tail lashing and hoofs kicking, an enraged aurochs rampages past trotting ponies. This lively scene is taken from a Lascaux mural of considerable size: The aurochs alone measures five and one-half feet. The enigmatic lattice symbol at the left might signify a gate toward which the beast is being stampeded. The artist has created the impression of having used a wide range of colors, perhaps by painting in black over an older red image.

A bold charcoal sketch from the Niaux Cave in the French Pyrenees shows a stocky, long-haired horse. Simple outline paintings dominated cave art; texture was provided, as here, by the rough infill that gives the impression of a shaggy winter coat.

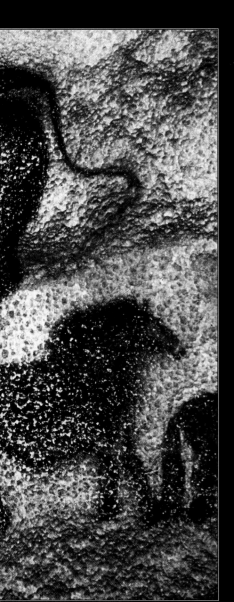

A bull charges across the rockface of a subterranean corridor at Lascaux. The difficulties of painting such a large portrait while maintaining accurate proportions were increased by choosing a site high up on the cave wall. Traces of sockets in the walls about six and one-half feet above ground level suggest that scaffolding—probably made of branches—supported a platform on which the painters could work. The artist here has twisted the perspective of the front hoof to show its underside, possibly to teach hunters how to recognize aurochs prints.

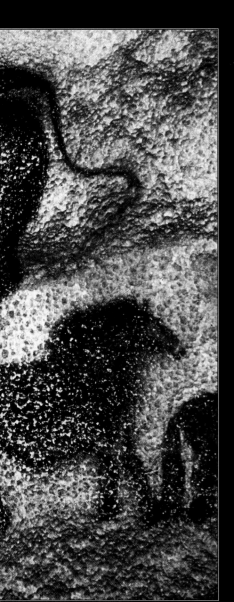

Natural indentations made by water droplets form an eye and a series of flank wounds in this ingenious clay-floor engraving of a wounded bison at Niaux. Most floor images in caves have been worn away by the passage of human feet; this one was protected by an overhanging rock.

Two sculpted clay bison, each about two feet long, lean against some rocks in a remote, low-ceilinged chamber of a Pyrenean cave. The animals were carved on the cave floor out of blocks of wet clay before being propped against the rockface for the fine detail to be added. Once completed, the sculptures were never visited, suggesting that the act of creating them was part of a private ritual.

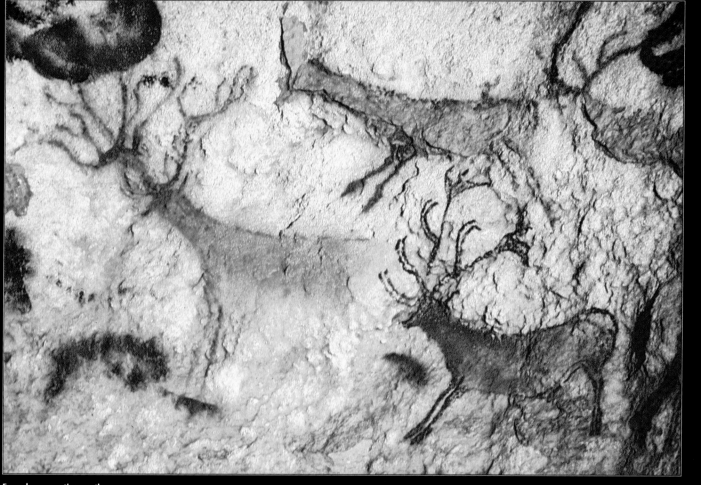

Four deer, sporting vastly exaggerated antlers, career across a wall of the Lascaux Cave. They are exceptional for being outlined in red and yellow as well as the more usual black. Cave artists manufactured their paints by grinding natural earth and rock pigment to a fine powder, then mixing it with water to produce a paste or liquid.

A head-to-head confrontation between two ibexes is the subject of this bas-relief limestone carving, found in a cliff-face recess in France's Charente. Most Ice Age bas-relief work is found in the open air, but the reason for this remains obscure; the high standard of cave painting indicates that adequate illumination existed inside the caves.

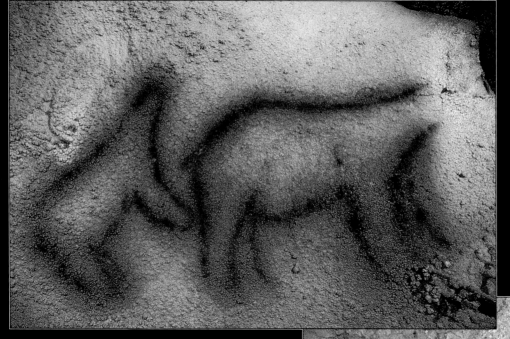

The image of a hair-coated mammoth *(right)* on a wall at Pech-Merle provides a reminder of the freezing temperatures that existed in southern Europe during the last Ice Age. The abbreviated lines of the drawing indicate that this work was something of a lightning sketch, probably taking less than five minutes to complete. It forms part of a large mural featuring twenty-five animals and measuring twenty-three by eight feet.

Unusual for being depicted fullface, the painted lion *(right)* peers from a wall at Les Trois Frères Cave in southern France. Ice Age artists rarely portrayed carnivorous beasts such as the lion or the two headless bears *(above)* from the Ekain Cave in northeast Spain, perhaps because close observation was too dangerous. Nocturnal creatures such as the owls *(below)* from Les Trois Frères are also infrequent, perhaps because of the lack of opportunity for studying them.

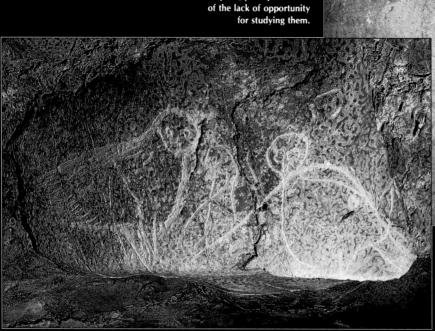

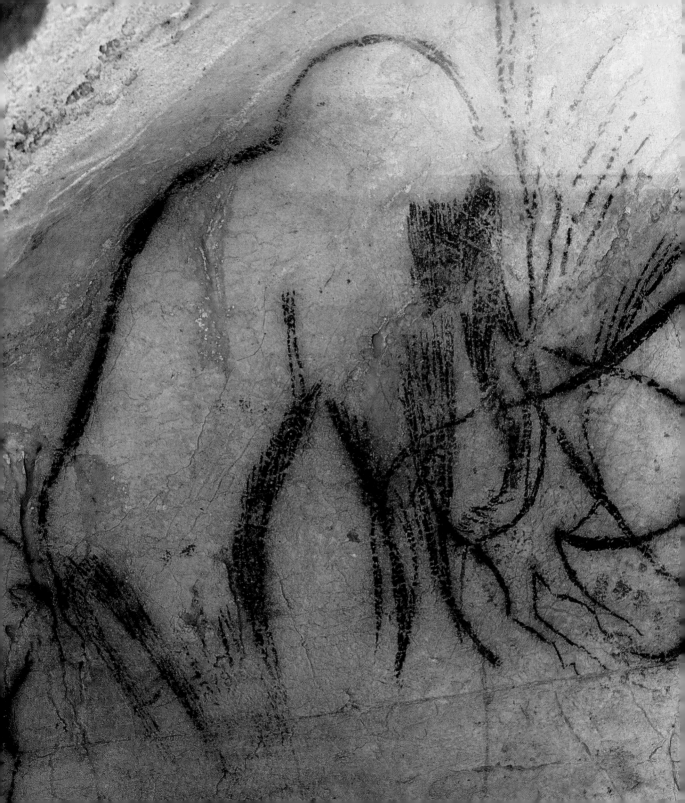

TAMING THE WILDERNESS

The camp at Beidha, in the Valley of the Ravens, had long been a favorite stopover for nomads wandering through the sandstone mountains of southern Jordan. It was a well-chosen site: Set on a wide terrace overlooking the wadi bed to the west, it afforded a good view of possible prey or enemies; water was available in nearby rock pools, supplied by a permanent spring that gushed from the crags 1,300 feet above the camp; and the low ground offered a plentiful supply of food plants and wild animals. Year after year, small bands of hunters would halt there, clearing away the undergrowth and firming up the soft, sandy ground with layers of clay before setting up temporary shelters. It was not worth constructing a more permanent base, for when the surrounding countryside had been scoured of all sustenance—a matter of perhaps a month—the hunters would strike camp and move off to new pastures. In one or two seasons' time, the game would return, the berries and plants would reappear, and the next group of visitors would arrive.

For centuries, Beidha had witnessed this transient passage of humanity. Then, shortly before 7000 BC, the rhythm of the valley began to change. For some reason the nomads who walked up to the terrace were staying longer. On the site of the earlier huts sprang clusters of round houses, whose floors were sunk eighteen inches into the ground. Above each pit rose a ring of sturdy wooden posts, enclosed by plastered stone walls and roofed with a clay-topped reed thatch. This was no temporary accommodation. The buildings' very structures were designed for year-round occupancy: The clay roofs kept off seasonal rains, while the sunken floors offered extra insulation during the cold winter months.

The new arrivals were too canny to think of themselves as permanent residents; there was always the possibility that drought, earthquake, or some other natural disaster would force them to move on again. Besides, they would have been uncomfortably aware of one major drawback to their otherwise sturdy dwellings: The timber framework around which each house was built burned well and frequently, which often resulted in wholesale devastation of the nascent village. But from just such a tragedy would emerge a clue as to the reason for this sudden change in the valley's transient population.

On one occasion, as the posts charred and snapped, the roof crashed down, and the walls tumbled in, a fleeing family discarded two precious baskets of food plants. Preserved beneath the debris, these baskets contained a small scrap of carbonized knowledge that, 9,000 years later, would reveal Beidha's secret. When the site was excavated and subjected to the spotlight of modern scientific scrutiny, it was discovered that the baskets had contained grains of wheat. That in itself was nothing unusual: Wheat grew—and still grows—as a wild grass in the region. But the grains varied in size and structure, indicating that important physiological changes were

A pottery figure made around 4500 BC clasps a broad-bladed farmer's sickle over its shoulder. Found at Szegvar-Tüzköves in eastern Hungary, this ten-inch-high statuette clearly portrays someone of bearing and importance; the regal, seated posture possibly indicates a deity. Agriculture came to Hungary in the sixth millennium BC, having spread through the Balkans from its origins in the Near East. By the beginning of the fifth millennium BC, farming had become established in Europe as far north as the Netherlands.

taking place. These changes were the result of the people of Beidha cultivating their own crops. Planted in the valley's fertile soil and watered by the nearby spring, the grain would have provided the Beidhans with a ready source of food: They would have had no need to desert the valley in search of a fresh food supply; it would have been worth their while to build permanent dwellings—and to rebuild them after each conflagration. Agriculture had arrived in the Valley of the Ravens.

For two million years, human beings had been on the move, living off what they could forage from the wild. Villages such as Beidha heralded the end of all that. For throughout the Middle East, similar settlements were springing up—and no doubt burning down—as other groups of hunter-gatherers strove to master the mechanics of farming: the sowing and harvesting of crops and the herding of domesticated livestock. These seemingly simple innovations were to change the face of the world. Within a few millennia, the disparate mass of humanity had begun to coalesce into sedentary clusters that dotted every veld, steppe, plain, and pampa on the globe.

It was not a sudden revolution. The habits of generations died hard, and initially, the old methods of hunting and gathering continued alongside the new way of life. Nevertheless, the pattern of human existence was irrevocably changing. For by bending nature to their will, the first farmers had done more than simply increase their food supply. They had given themselves at least a chance—barring crop

The map at the bottom of the page shows the locations of the earliest-known domestic crops and the directions in which initial cultivation of those crops is thought to have spread. Much early crop domestication took place in the area of land known today as the Fertile Crescent, shaded green on the inset map. This region owed its fertility to a warm climate and to the annual winter rains. The resulting abundance of plants and wildlife enabled hunter-gatherers to abandon their nomadic lifestyle and to establish settled farming communities, as shown in the enlarged area.

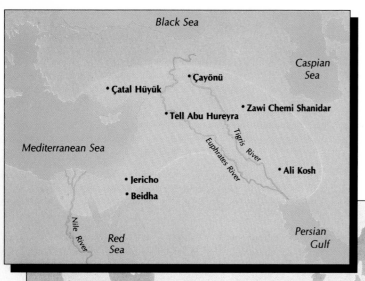

failure and disease—of living in greater security and to a greater age. They had the option to diversify, to make and trade desirable possessions, to accumulate wealth, and to improve—indeed, to create—their lot. In short, they had gained some measure of control over their destinies, albeit they had, in the process, lost some of the social and economic flexibility that had accompanied their old nomadic lifestyle. Once they had had a taste of this stability, few could resist it, and the settled, agricultural population steadily grew. Small villages became towns, towns expanded into cities, and these, in turn, became administrative centers for territories that would ultimately become empires. The foundations had been laid for civilization as we know it today.

Twelve thousand years ago, as the frozen plateaus of the Ice Age shrank back to the North and South poles, the world again came to life. Released from the climatic pall of the Ice Age, vegetation spread, animals abounded, and humans multiplied. By 10,000 BC, there were between 5 and 10 million people on the earth, most of whom lived by their well-developed skills in hunting and gathering. They scouted in small groups across the countryside, settling where the food was sufficient to supply their daily needs and moving on when sustenance became scarce. They were well aware of the natural cycle of reproduction, growth, and death and also of the changing seasons. They knew their landscape well—where best to find grain seeds, roots, and fruit, and where to hunt game at various times of the year. Material wealth was limited to what they could carry, but provided the group was small enough, life could still be good: The meat they acquired and their main staple of edible vegetation provided a balanced and nutritious diet; they could grind grain into flour to make bread; and they knew which plants could be used to heal wounds and cure illnesses.

The thawing of the icecaps, however, had disrupted the established order. For in spite of the hunter-gatherers' skills and intimate knowledge of the ways of nature, their way of life best suited only relatively small bands of about twenty-five to thirty people. As a group grew in size, it became more difficult both to manage and to feed. When the numbers became too large, family units were forced to split off to discover new hunting territory—usually within the same region. As long as the population remained relatively small and land was freely available, this dispersion presented no problem. But with the earth's new fruitfulness, humanity was on the increase. In addition, huge areas that had previously been dry land—such as large parts of the Mediterranean and Black seas, and later the North Sea—were now submerged by the rising sea level, and their inhabitants were forced inland to find new areas to exploit. As a growing number of people competed for the richest and most accessible ground, it became increasingly necessary to consider a different form of subsistence strategy.

The first to do so were the hunter-gatherers of the Middle East. There, moist winds filtered eastward from the Mediterranean Sea to reach the Fertile Crescent, a strip of hilly but productive land that arched north from Jordan to the foothills of Asia Minor before dropping down along the Tigris and Euphrates rivers in Mesopotamia to reach the Persian Gulf. It was good hunting ground: On the upper ridges, wild goats and ibexes found a favorable foothold; while on the lower slopes down to the valley floors, sheep and gazelles grazed alongside wild cattle and asses. High on the hillsides, open woodland flourished with oak, juniper, hawthorn, pistachio, and wild pear trees. More important, however, in those areas favored by the sun and seasonal rains, vast fields of wild cereal grasses covered the ground, in some places spreading over thousands of acres.

It was an abundance too great to pass by. Small bands of nomads would stop to set up camp beneath an overhanging cliff, in a suitable cave, or under rough shelters made of branches covered with leaves and grass. And as it gradually became clear that there was enough food for at least semipermanent occupation, so these rough camps became more substantial collections of huts. The densely packed fields of wild cereal—barley and two different strains of wheat, emmer and einkorn—provided ample food. The wheat was of high quality, containing 50 percent more protein than that used for bread flour today, and each acre yielded some 700 pounds of consumable grain per year. During the three-week harvesting season, a family, stripping the stalks with their calloused hands or with flint-bladed reaping knives, could gather enough grain to meet their annual needs and still have some to spare.

There was little chance that this bounty would diminish, for the wild cereals proliferated with tenacious efficiency. Each seed was covered by a sharp-pointed husk along whose sides ran a series of backward-sweeping hairs, terminating in a few long bristles. As the ears ripened in the late spring, they shattered, throwing their aerodynamic burden to the wind. The seeds scattered like darts, penetrating the soil, point first, where their hairs held them in position; any disturbance by wind or rain simply made them wriggle more deeply into the ground. During the hot, dry summer, the seeds lay dormant, until the winter rains came to flush them farther into the soil to germinate and to grow the next season.

Nevertheless, the evolutionary characteristics that had produced such plenty also militated against those who wished to enjoy it. Harvesting this brittle and elusive crop was an infuriating task. Because the ears of wheat and barley ripened quickly in order to escape the unwelcome attention of predators, timing was of the essence: Too early a trip to any particular wheat field resulted in an easily gathered but unripe yield; too late a sortie might find the harvester standing frustrated amid a sea of seedless stalks; and even if the moment were perfect, the slightest touch could send the precious grain scattering to the ground, from where its retrieval was a backbreaking business. Still more irritating was the fact that the individual seeds of each ear ripened at different times, so every successful haul was likely to contain equal quantities of good and bad grain. Among the grasses, however, there were some of a mutant strain

Four maps show where livestock was first domesticated and pinpoint sites where the earliest evidence of this has been found. These animals provided human beings with meat, milk, and later, hauling power, but they belonged to wilder, usually larger breeds than their modern-day descendants.

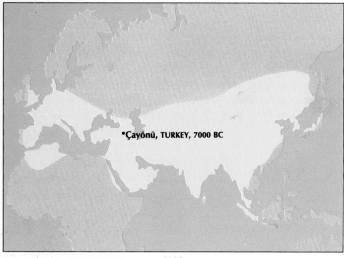

•Çayönü, TURKEY, 7000 BC

PIGS

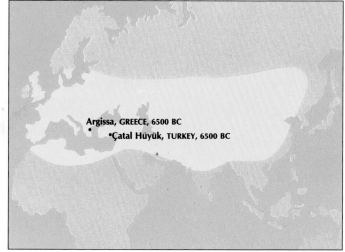

Argissa, GREECE, 6500 BC
•
•Çatal Hüyük, TURKEY, 6500 BC

CATTLE

whose seeds not only ripened at the same time but also remained on the plant until the whole ear dropped to the ground. Standing out like beacons of plenty against the spindly outlines of their more dynamic cousins, these easily gathered, plump specimens would have been the first to be carried home in the harvesters' baskets.

Back at the camp, the day's yield would undergo the necessary processing to make it fit for consumption. First, the seeds were threshed with sticks to remove the rough outer casings; then, the resulting mixture would be tossed into the air to allow the light chaff to blow away, leaving only the nutritious kernels; next, a sifting process would remove the smaller weed seeds; and finally, the end product would be ground between stones in preparation for cooking. Amid this flurry of activity, some of the seeds would inevitably be spilled and swept onto the communal garbage dump where they thrived untended. Eventually, these accidents of replanting grew in number to form a new stand, located not out in the open countryside but on the outskirts of the settlement itself, an arrangement that was both more convenient for gathering and easier to protect against scavengers. Almost unconsciously, the first steps had been taken toward crop cultivation.

It was a short step from opportunism to intent. Soon people began putting aside some of the precious nonshattering seeds, storing them either in straw-lined pits in the ground or in pottery jars outside their homes, and then sowing them the following season to ensure another harvest of easily gathered grain. And gradually, over many years, farmers picked out and nurtured other mutant crops, thereby producing a number of changes in the makeup and character of the seed, among them an increase in both the size and number of grains on each ear. Despite the advantages of these new domesticated strains, wild cereals continued to be harvested; their seeds' hard outer coating made it more difficult for birds to steal the grain in the fields and also gave the seeds a longer storage life. All the same, the mutant grasses' greater productivity could not be ignored, and steadily, the settled population began to expand.

Not only were the economics of the situation becoming increasingly obvious— about 250 square miles of land were needed to sustain 25 hunter-gatherers, while a mere 5.8 square miles could feed a community of 150 farmers—but village life also offered other advantages. There were fewer hunting accidents, for instance, and

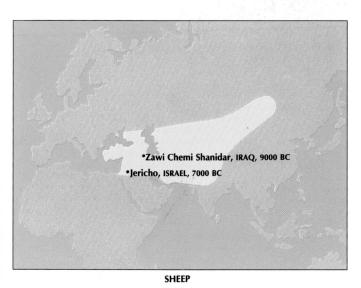

•Zawi Chemi Shanidar, IRAQ, 9000 BC

•Jericho, ISRAEL, 7000 BC

SHEEP

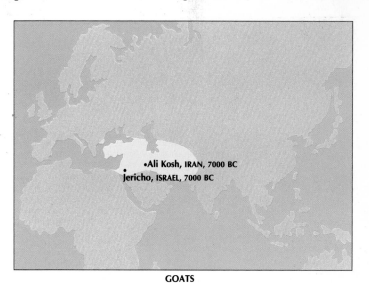

•Ali Kosh, IRAN, 7000 BC

•Jericho, ISRAEL, 7000 BC

GOATS

97

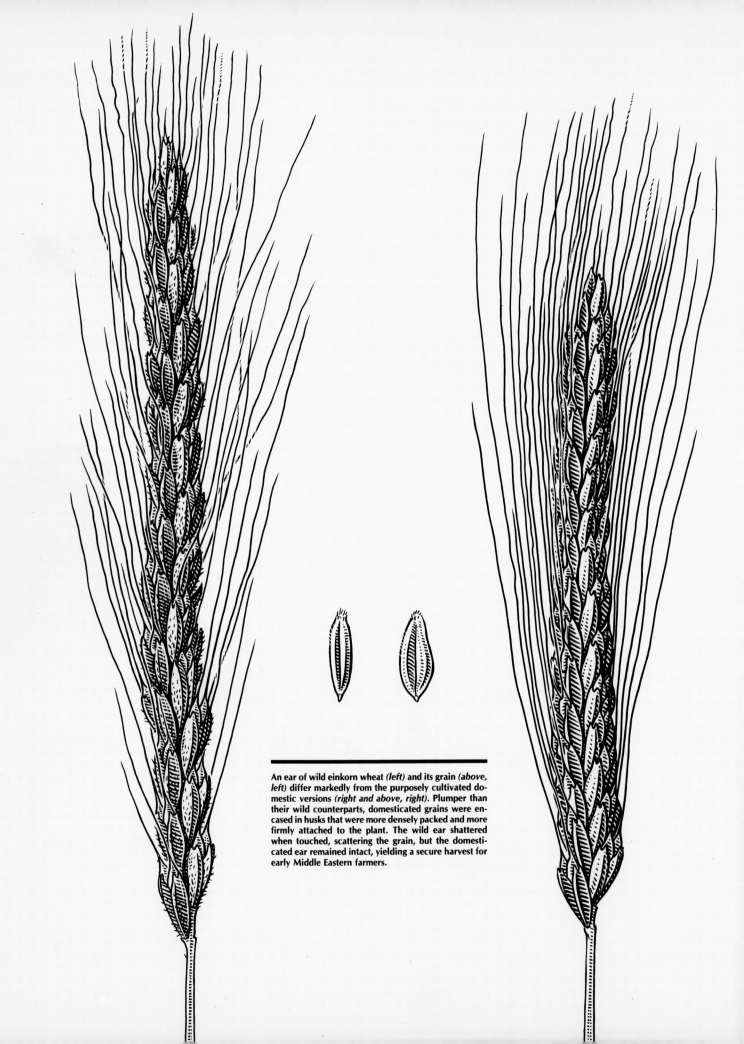

An ear of wild einkorn wheat *(left)* and its grain *(above, left)* differ markedly from the purposely cultivated domestic versions *(right and above, right)*. Plumper than their wild counterparts, domesticated grains were encased in husks that were more densely packed and more firmly attached to the plant. The wild ear shattered when touched, scattering the grain, but the domesticated ear remained intact, yielding a secure harvest for early Middle Eastern farmers.

greater protection against carnivorous predators, such as lions and leopards. Whereas small children had previously been a liability on the trail, it was now possible to raise larger families without fear of hampering the rest of the group. And the elderly, who had often been an encumbrance, could now take a place in society, maintaining a continuity of experience, skills, and wisdom. It was the older members of the community, too, who would look after young children, freeing the parents to work in the fields or to go hunting.

As the sufficiency of food turned into a surplus, so the settlers had both the spare time and the inclination to manufacture the kind of desirable possessions that during their nomadic days had been too much trouble to carry around. As a raw material—especially in areas that lacked suitable flint for making tools—they particularly favored obsidian, a shiny, black volcanic glass that was crucial to hunters and farmers alike. From a suitable chunk of this material, an artisan would skillfully chip off razor-sharp flakes for arrowheads, which were sometimes embellished with a barb and bound to the shaft with fiber, or glued in position with gum from an aromatic mastic tree. Smaller chips could be set into a curved piece of wood to form a sickle, while large flakes could be shaped to form knives and scrapers that the settlers used for cleaning meat from hides.

Nowhere, perhaps, were the advantages of settled life enjoyed more than at the large center of Jericho. Situated by a perennial spring that pumped almost 1,200 gallons of fresh water a day into the semidesert of the lower Jordan Valley, Jericho had witnessed one of the earliest instances of crop cultivation. As early as 8000 BC, the surrounding fields were yielding a double and sometimes triple crop of wheat and barley, which fed a population of perhaps as many as 3,000 people. At the same time, Jericho also supported the many peasants in the surrounding area, who came to the center to barter obsidian, salt, semiprecious stones, and luxuries such as cowrie shells and green malachite that could be fashioned into jewelry.

Long-distance traders had been operating for many years before the advent of farming; since 9000 BC, they had bought and sold obsidian along a well-established trade route that reached out from the volcanic areas of Acigol and Nemrut Dağ in Asia Minor to encompass the entire Fertile Crescent. Grain, which was heavy, though otherwise easy to transport in large quantities, may well have been one of the commodities in which they did business, and it is possible that these early merchants helped propagate the new, agricultural way of life.

At the same time, however, the expanding farming population was unable to sustain itself. Not all settlements were as fortunately situated as that at Jericho. Most had limited resources, and as populations grew and villages neared the starvation point, a small band—such as that which arrived at Beidha—would gather up a few baskets of precious grain and move off to new territory.

The settlers at Beidha were just one small part of an economic migration that, by 7000 BC, had spread the process of selecting and cultivating grain throughout the Fertile Crescent. And it was not long before other plants were brought into cultivation. The wild, pod-bearing pea, for example, grew as a prolific weed, invading the grain fields and, by the fourth millennium BC, becoming a common source of protein. Over the years, from the large number of wild plants that had filled humanity's cooking pots for thousands of years, a few were gradually selected for cultivation—ancestors of familiar foods such as carrots and cabbages, onions, olives, pears, and plums.

In addition, the settlers soon discovered that plants could yield more than just food.

In the seventh millennium BC, farmers in the Fertile Crescent cultivated flax, not only for its linseed oil but also for its fiber.

Producing cloth from vegetable matter was a tedious process. First, the flax stems had to be soaked, then vigorously pounded to release the fibers. Next, the fibers were spun—either by rubbing them on the thigh or by twisting them on a free-hanging spindle—before being woven on a horizontal loom that was staked out on the ground. The finished product, however, was well worth all the trouble—a fine linen that was both lighter and more comfortable than the skins and woven animal hair previously worn by settlers.

A further incentive to develop crop domestication was the discovery of fermentation. Probably some happy accident in the processing of barley grains revealed the secret of malting, and the potion it produced must have introduced a magical dimension to village life. Later, inhabitants of the Mesopotamian region of Sumer enjoyed at least eight barley beers, eight emmer beers, and three mixed brews. The beverages were not always the purest: Sometimes a straw was required to get beneath the layer of surface detritus. Nevertheless, the Sumerians rated the product highly enough to give it its own goddess, Ninkasi, "the lady who fills the mouth."

Fruits, too, were cultivated for alcohol, but not before a great deal of trial and error and the discovery that a sweetener such as honey or nectar was needed to start fermentation. Regular cultivation of grapes, perhaps for winemaking, was being carried on in the fifth millennium BC down the eastern flank of the Fertile Crescent. Vineyards in southern Mesopotamia were exporting wine to Egypt some time before 3000 BC, and at the time of Abraham, around 1800 BC, it is recorded that vineyards near Haran in northern Mesopotamia were producing wine from 29,000 vines.

Despite the joys of alcohol, the new way of life did not endear itself to everyone. Working the land was hard, especially at the beginning of the growing season. First, the farmers had to till the ground with digging sticks—slender poles, about three feet in length, whose hafts were sometimes weighted with stone disks for added force. Then, they had to burn the residue from last season's harvest before the new crop could be sown; afterward, as the grain sprouted, unwanted weeds had to be either pulled out or removed by hoe, and the field had to be guarded against predators until the crop was ready to be gathered in. Only when food resources became scarce would people turn to this tiring existence, and for a long time, crop cultivation merely supplemented the old methods of hunting and gathering.

The long-lasting importance of the hunting and gathering tradition can be seen at the village of Abu Hureyra, set on the banks of the Euphrates River in northern Syria. The Hureyrans were agricultural pioneers, having domesticated cereal crops at least as early as did the inhabitants of Jericho. Nevertheless, for more than 2,000 years, they continued to rely heavily on hunting. Meat was plentiful in the surrounding area; onager—a species similar to the wild ass—goats, sheep, pigs, and wild cattle roamed the steppes above the river valley, and every year, the region was visited by gazelles, who congregated on the banks of the Euphrates during the long summer drought. So rich a food source was the gazelle that, by 8000 BC, Abu Hureyra was home to a community of several hundred (or perhaps several thousand) people, who derived 80 percent of their meat requirements from this animal.

The slaughter of a herd of these agile, light-footed creatures required a special strategy. Large-scale traps were constructed, usually near a valley through which gazelles were known to move. This type of trap was a wedge formed by two stone walls, set well apart at one end, that gradually converged over a distance of several miles until they debouched into a circular, walled enclosure up to 500 feet in diameter. A herd, migrating through the valley, was stampeded and driven down this narrowing corridor into the enclosure. The entrance was then closed off, and from concealed niches around the walls, hunters would slaughter the trapped animals with a hail of arrows and spears. As an alternative, deep pits were dug behind specially lowered sections of wall. The gazelles leaped the wall, then fell into the pits, where they were quickly dispatched. The carcasses were processed at the killing site, the long bones being discarded and the meat being salted and dried for storage and consumption throughout the year.

So great was the migratory passage of game that huge numbers of these rudimentary abattoirs were constructed, running south for hundreds of miles across the killing fields of Syria. Some were linked in overlapping clusters to form chains that took in whole herds at a time, and the great annual slaughter continued apace. But by about 6300 BC, the devastation was beginning to upset the gazelles' instinctive pattern of migration and was even threatening the very survival of the species. As their main supply of food dwindled, the people of Abu Hureyra were forced to fall back on a different pattern of subsistence.

By then, however, the early humans' knowledge of the animal world had been put to a less wanton use. As permanent settlements became more widespread, their inhabitants sought to apply the benefits of domestication to the animals they hunted—the most notable of which were goats and sheep. These two species were ubiquitous in the Middle East, thriving, respectively, in the mountain uplands and the rolling foothills, wherever there were vegetation and water. The more daring members of a wild herd might even be seen within a village, snatching discarded food and browsing on garbage dumps.

Two traits in particular make goats and sheep ideal candidates for domestication.

A primitive stone mortar and pestle found in Iraq—dated between 7000 and 4000 BC—provides evidence that the first farmers of the Middle East developed techniques for turning wheat and barley grains into digestible forms of food. They used the grinding stone to break open the inner husks that contained the grain. Once freed, the grain could be winnowed to remove chaff, then either mixed with water to make gruel or, after more grinding, used to make unleavened bread.

Both are herd animals, accustomed to staying in groups with a recognized leader, and both form quick attachments to their parents at an early age. A young kid or lamb, separated from the herd—and probably suckled by a human wet nurse—would have bonded with its human parent. When it grew to maturity and bred, it would then provide part of a tame and virtually self-supporting flock. In the wild, these animals represented food, but in captivity, they provided wool for clothing and dung for fuel and fertilizer. Moreover, they increased farm output by being able to graze on marginal land that was unsuitable for cultivation. And after the mechanics of milking were mastered, the early farmers' diet was augmented by butter and cheese.

As they had adapted cereals to their needs, the settlers were quick to adjust this new evolution to meet their own ends. Within a few generations, goats and sheep had been bred to become smaller and more docile. Even their horns became rounder and less dangerous—or, in some cases, they disappeared entirely. Productivity, too, was boosted. Wild sheep, for example, which had come under human control as early as 9000 BC, originally possessed a stiff, hairy coat with a woolly undercoat that was shed annually. In time, domestication and crossbreeding produced an animal whose woolen fleece—in various shades and combinations of black, brown, and white—was not shed each year but was sheared for the well-being of both human beings, who acquired a welcome extra layer of winter clothing, and beasts, which got rid of an unwanted summer overcoat.

The beginning of the sixth millennium BC saw flocks of sheep and goats roaming at will through virtually every settlement in the Middle East. And by this time, too, the barnyard clamor had been swelled by the arrival of the pig. Fleshy, fast-breeding, and capable of eating almost anything, the pig was an invaluable meat factory. When penned within an enclosed space, it was prodigiously efficient at turning the soil, and such was its intelligence that, when turned loose in the surrounding countryside during the day, it would return of its own accord to its owner at night.

By 6000 BC, in the comparatively sophisticated center of Çatal Hüyük, in southern Anatolia, the domestication of wild animals had broadened to include a less likely candidate: the aurochs, progenitor of modern cattle. Standing six feet tall at the shoulder, with long, sharp horns, heavy muscles, and an aggressive disposition, these were formidable beasts. That they came to be domesticated is a testament to human resolve and ingenuity. The male aurochs would have been very difficult to restrain, and it is unlikely that the smaller, but by no means docile, female would have allowed herself to be milked. They were difficult to manage and had to be kept away from crop fields, which could easily be trampled and destroyed. But almost certainly, there were religious as well as agricultural reasons for their capture. Among the many animals in the wild, the aurochs was a clear example of powerful strength and fertility. Its crescent-shaped horns were associated with the moon—which had a special place in the religious rituals of the people—and it may have been captured at first for sacrificial purposes.

The braver farmers would have laid traps—probably leg snares—for passing aurochs. Those who did not relish the prospect of trying to subdue a snared and no doubt furious beast would have chosen a less confrontational method. Their first step would have been to leave water and lumps of salt near a settlement. As the animals became familiar with both the site and the presence of humans, they would have become more approachable. And from approachability, it was but a short step to capture. The more amenable stock would be selected and encouraged to breed with the more docile cows to produce a higher milk yield, while bulls not kept for breeding would be castrated to make them less ferocious. In the process, selective breeding modified the characteristics of the species, reducing its size and, in many cases, changing the shape of its horns. Indeed, so great was the early stockbreeders' ardor to create a manageable beast that by the first millennium BC, the shoulders of some cattle were no higher than three feet off the ground.

In addition to rearing animals for food, humans were also learning how to make beasts work for them. Probably the early hunter-gatherers had struck up a semidomestic relationship with the wolf—to whom all present-day dogs owe their ancestry. A sharp-toothed and resourceful beast, who competed directly with humans for available game, the wolf was a menace in the wild, but an invaluable ally when trained. As animals that hunted in packs, shared their food, and recognized the authority of a single leader, wolves were remarkably similar to humans, and when captured and reared from cubs, they readily accepted life under their new masters. Although they were used initially by hunters to flush game, wolves soon became protectors and herders of livestock. Closer to home, they also acted as an efficient early-warning system against the approach of strangers and could even be used as pack animals to haul light loads.

It was longer, however, before other animals were pressed into the service of humans. Not until at least 4000 BC or later were the horse, camel, and ass tamed to provide a means of transportation for the settlers; and a millennium later, cats came off the trash dumps to take on the role of household ratcatchers.

Once people had become settled in one place for long periods of time, they began to build more substantial houses for themselves and their growing families. By 6000 BC, small townships in central Mesopotamia had expanded to

Carved on a lapis-lazuli cylinder seal *(left)*, a Sumerian drinker of the third millennium BC quaffs beer from a deep bowl. To reach through the scum that formed on the beer's surface during brewing, he uses a long straw, although it may not have been as precious an item as the gold straw shown here, which measures four and one-half feet long and which belonged to a high-ranking priestess in the temple of the moon god Nanna, at Ur. A drawing showing more of the seal *(above)* indicates that beer drinking was a communal affair.

103

cover areas of ten to twelve acres with houses of mud-brick walls and roofs made of timber and plaster. Each dwelling had a hearth for cooking the daily meal—a thin gruel or a meat stew, supplemented by boiled roots, bread, fruit, and nuts and possibly a thick porridge of wheat grain. The hearth, too, provided warmth for the family, as they huddled under their animal skins during the cold winter nights.

Beside some of the houses stood outbuildings used for long-term grain storage. To others were added small rooms where settlers could make weapons, tools, and personal adornments for their own use or for barter. Beidha, for example, boasted a twenty-foot-long row of workshops that provided workspace and upper-floor living quarters for six artisans, who specialized in different materials: A boneworker might make objects such as pins, beads, necklaces, or shoulderbone shovels; stoneworkers would produce agricultural tools, grinding stones, and mortars and pestles. In Beidha, there were even butcher and beadmaker shops.

The most ubiquitous craft, however, was that of the potter. Clay vessels had been constructed by Japanese hunters and gatherers as early as the eleventh millennium BC, but their very nature—both fragile and heavy—made them impractical. As a more permanent way of life was adopted in the Near East, pottery jars became essential equipment for storage and cooking. The techniques used for their manufacture were simple, requiring neither special skills nor tools, and virtually all members of society could make their own pots. Containers of all shapes and sizes were fashioned by rolling long ropes of clay that were then curled around, one layer on another, to whatever height was required. Others were made either by kneading a ball of clay into the desired shape with the bare hands or by forming it around a mold of stone or wood. The vessel was then smoothed by hand before being baked on an open fire at a temperature of 1,500° F. to 1,600° F. By the fourth millennium BC, the introduction of the potter's wheel and high-temperature kilns had made pottery making a specialized occupation, requiring just a few skilled artisans to supply the needs of a whole community. But even before that time, the craft had developed beyond the simply utilitarian and had become a means whereby villagers could give their artistic expression full rein. Pigments were made from substances such as ocher and malachite to brighten the basic pottery vessels. And decorative elements were incorporated into the design, often inspired by familiar objects such as animals, human beings engaged in various activities, and religious symbols either to appease the gods or to ward off evil spirits. Toys were also constructed for the amusement of children: bracelets, bangles, necklaces, and models of people and animals.

Such modeling expertise was also pressed into the service of religion. There is little doubt that early humans' intimate association with nature led them to believe in and respect some external, invisible power that regulated the whole cycle of life: its growth, reproduction, and death. Many thousands of small, handmade clay figures were produced, representing symbols of fertility: naked female figures, enormously fat with exaggerated breasts and thighs, some in the position of childbirth.

In Mesopotamia, collective worship became institutionalized for the first time shortly after 5000 BC, in the town of Eridu, southwest of Ur. In honor of their patron deity Enki, who presided over the town's water supply, the people constructed a large building that served as both a temple and a focal point for the town's agricultural and manufacturing activities. Over the years, as each successive building fell into disrepair, another structure was erected upon its foundation, time and again, one upon another, in its later stages taking the form of a high, stepped tower, or ziggurat. This architectural form was to be repeated many years later in other parts of the world.

The temple was the hub of the town's life, where surplus agricultural and hand-crafted goods were collected and distributed under the auspices of the incumbent god. In the bustling marketplace, packed with sellers whose vegetables and hand-made wares lay spread out in front of them, the gossip and news that passed from one person to another were as important and valuable as the items that were traded between them. Along with the exchange of knowledge, however, the townspeople's conversation must occasionally have turned to darker matters: the price in death, destruction, and disaster that the early farmers paid for their success.

For the new way of life that had brought so many benefits and solved so many problems had its disadvantages, too. The stability of settled existence was constantly under threat from the vagaries of nature, such as epidemics of disease, flash floods, violent dust storms, scorching winds, and insect plagues. Moreover, it was not only natural hazards that beset early farmers. The accumulated wealth of a community was an obvious target for marauding groups of nomads or envious neighbors, and increasingly, the spears, slings, and arrows that had once been used against animals were now directed toward humans. Designed as it was to deal with such powerful creatures as the aurochs, the hunter's arsenal was devastatingly effective. A sling, for example, had a range of up to 200 yards and could launch a fist-size stone, or hardened-clay ball, with enough force to smash bones. If they did not post a guard, the inhabitants of an unprotected village were easy prey: All that night-raiders needed to do was stand outside the door of each house and slaughter the settlers, one by one, as they emerged. Accordingly, one of the first priorities for a new settlement was to build a protective enclosure.

Health was another problem facing the villagers. As settled farmers, producing relatively few types of food and possessing less time to forage for the wide variety of natural foods that the human body requires, they inevitably suffered nutritional-deficiency diseases. In addition, the accumulated filth of human settlement brought the bane—unknown to nomads—of food-contaminating vermin, such as rats and cockroaches. Constant, close contact with livestock produced new ailments: Measles, influenza, and smallpox were just a few of the viruses transferred, in one form or another, from animals to humans. Within larger settlements, a high density of people living in close proximity and having limited sanitation would have been prey to the spread of many infectious diseases such as dysentery, typhoid, and tuberculosis—a situation which was often aggravated by a settlement's geographical location. At Çatal Hüyük, for example, situated in a marshy lowland basin, malaria was rife and may have been the cause of many early deaths. Very few farming people in 6000 BC lived to middle age—thirty years for women and thirty-five years for men was the general life span that could reasonably be expected.

In spite of all the drawbacks of settled life, populations expanded rapidly, and the need to grow more food soon forced farmers to increase the amount of land under cultivation. One way to do this was by irrigation. It would hardly have escaped the attention of early settlers that the most productive tracts of land were those that were soaked each time the streams or rivers flooded. From that knowledge, it would have been a short step to devise a means to dam the river flow and divert the water through a ditch to reach drier land. This proved particularly fruitful in the plains of Mesopotamia, where the floods of the Tigris and Euphrates rivers laid down fertile deposits of alluvial silt in areas where the natural rainfall was insufficient to support crops.

It was easy enough in principle, albeit laborious in practice, to construct the irrigation system. Where the rivers flowed across flat plains, years of flooding and the continual depositing of silt had caused the water to run, within raised levees, above the level of the surrounding countryside. It required only a primitive shovel to cut through the riverbanks and form the ditches, plus a basket to carry away the earth. Each ditch that sprouted from the river was, in turn, connected to numerous minor channels, so that eventually, the ground was laced with an intricate web of waterways. Although simple to build, such a network demanded continual attention. The ditches had to be cleared regularly of their accumulated silt, and boulders or wooden boards had to be positioned to divert water through the various channels.

The villagers learned not only how to bring life to previously barren land but also how to intensify cultivation on existing fields. The digging sticks that early settlers had inherited from their nomadic ancestors were soon replaced by a more effective device, formed from a J-shape piece of wood or sapling, known today as a scratch plow. To use it, one farmer pulled on a length of rope that was attached to the longer end of the piece of wood, while another plunged the shorter section into the earth, directing it across the field to form a shallow furrow.

It took longer, however, for the early farming communities to recognize the full potential of the powerful cattle that had taken them so much courage and effort to domesticate. Previously, they may have harnessed the strength of oxen to pull large boulders from their fields or to shift fallen trees, but it was not until the middle of the fifth millennium BC that agriculture experienced a significant change, brought about by the application of animal strength. With an amenable, trained beast harnessed to a plow, farmers could not only till fields far more quickly than before but could also open up ground that was too heavy or too weed-choked to be cleared by hand.

The first plow was a rudimentary affair: a single, stout stick, pulled by a rope attached to an ox, that scratched shallow seed-rows in the soil. It may not have been until the third millennium BC that farmers in Egypt and Mesopotamia produced a more sophisticated design: a heavy piece of pointed wood to penetrate the soil, with a crossbeam connected to the ox by ropes; two handles for steering the plow in a straight line; and a funneled tube immediately above the plowshare that allowed the farmer to pour seed directly into the furrow, treading it in as he went along.

Other methods of increasing food production were spurred more by necessity than by design. Year after year, the early farmers planted their grain, and inevitably, the soil became increasingly impoverished. They may not have understood why the annual crop was diminishing—that the nitrogen, phosphates, and potassium required by growing vegetation were becoming exhausted. For whatever reason, the early farmers would often abandon the stricken field and clear a new space for cultivation. Livestock were turned loose onto a piece of nearby open woodland to crop as much of the standing vegetation as possible, while at the same time nourishing the soil with droppings. Trees were felled with stone axes or killed by ringbarking—cutting around the trunk's circumference to stop the flow of sap—before being cut into manageable pieces and taken to the village for firewood. Reaping knives dealt with bushes and shrubs, while digging sticks were employed to pry out roots and boulders. The work was done in the hot, dry months of the year, before the rain came and the seed had to be planted. Farmers burned the timber and debris that had dried in the hot sun, leaving ash and dying vegetation to enrich the soil. Cattle were again driven over the clearing to trample the ground as the next season's seeds were scattered. The new crop would be carefully watched as it grew—after the sweat and toil of so many people, it could not be allowed to be demolished by stray animals or choked by unwanted vegetation. The new season's harvest, rich and plentiful, more than paid for their united efforts. Meanwhile, the old, abandoned field was ignored as it slowly reverted to its natural state, steadily regaining its fertility by means of decaying plants and rainborne salts until it was ready for use once more.

As long as there was enough fresh land to cultivate to keep pace with the increasing size of the community, this cycle of abandonment and expansion was effective. But in areas restricted by unsuitable growing terrain, enclosed by hills, or hampered by poor water drainage, the system had serious limitations. It took twenty to twenty-five years of lying fallow before an abandoned field could produce another healthy crop. And in many settlements where land was scarce, the fallow fields had to be brought back into use before the necessary nutrients had been restored. To compensate for lower crop yields, the villagers increased the quantity of their livestock, often above the limit that the available grazing could safely sustain. Overgrazing, continuous cultivation of the same fields, and possibly a few seasons of drought led to the desertification of once-luxuriant regions. It was enough to force the villagers to abandon their settlement and

VESSELS FOR THE HOUSEHOLD

The making of pottery containers came only with the development of settled communities; heavy and breakable objects were of no use to earlier nomads. The first-known pots were made in Japan around 10,500 BC: coils or lumps of clay pressed together by hand and either hardened over a bonfire or baked in a covered pit. By 3500 BC, Mesopotamian potters were using stone or wooden turntables to rotate their pots during the molding process—as depicted in a limestone statuette of a working potter *(right)*, found in Egypt and dated between 2500 and 2200 BC—and specially built kilns to fire them. Center shafts later enabled the turntables to spin more quickly, thus speeding up the production process. Examples on the following pages chart the progress of the early potter's art.

to seek out a new location, as their ancestors had done so many centuries before.

Farming soon spread westward to Europe. Here, the waning of the Ice Age had brought problems as well as plenty to the local hunter-gatherers. Previously, much of the landmass had been a cool steppe, home to herds of horses, deer, and bison. The existence of the giant Irish deer, which boasted an antler span of as much as twelve feet, was just one testimony to the lushness of the grazing. As the glaciers retreated, however, Europe was covered by woodland—first, by hardy specimens such as pine and birch and, later, by the warmth-loving oak and beech. Within these forests, hunters could no longer find whole herds; they were forced instead to dart between the trees, killing lone beasts with spears and arrows. To the north, a few hardy hunters still clung to tradition, culling the reindeer during their annual migration, or skimming across the frozen wastes on skis and dog-drawn sleds in pursuit of herds of elk. In general, however, the best living was now made by fishermen, who paddled their skin-covered coracles across the continent's rivers and coastal estuaries.

The new farming techniques from the Middle East wrought transformations on this timbered landscape, rapid in some areas, gradual in others. By 6500 BC, agriculture had come to the Balkans; 500 years later, it was established in Italy; and another

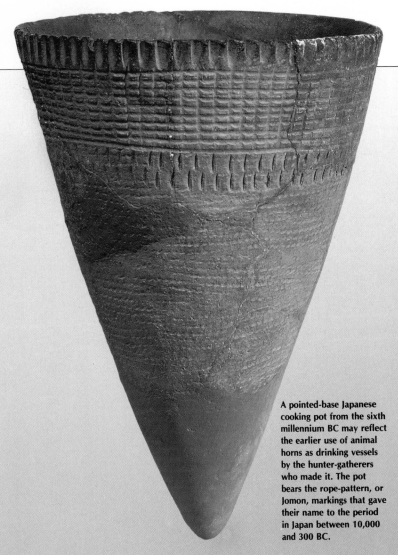

A pointed-base Japanese cooking pot from the sixth millennium BC may reflect the earlier use of animal horns as drinking vessels by the hunter-gatherers who made it. The pot bears the rope-pattern, or Jomon, markings that gave their name to the period in Japan between 10,000 and 300 BC.

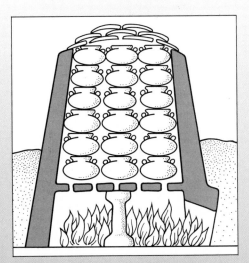

Clay pots are fired inside an early Near Eastern ceramic kiln, shown here in a cross-section diagram. A ventilated clay floor separated the pots from the flames, making cracking less likely. On top of the kiln, loose shards provided insulation; a draft tunnel at the base allowed for temperature control.

millennium or so saw settlements springing up in eastern and southern France. It took time to develop new strains of domesticated grain that could thrive in the cooler and moister conditions of northern Europe. But by 4000 BC, successful cultivation found its way northward through Hungary into the Netherlands and Poland.

At times, the dense forests that abounded were a hindrance. They made travel slow and grazing difficult. Settlers were required to cut down entire tracts of woodland in order to make clearings in which to pasture their sheep and goats. At other times, the trees were a boon: Not only did they prevent soil erosion but they also provided plentiful building material. In woodland clearings and along the river basins whose rich loess, or alluvial silt, provided fertile ground, Europe's farmers constructed wooden settlements. Poles were sunk into the ground to form the frames of thatched houses, as much as 130 feet in length, whose walls were composed of wattle and daub—thin, interwoven branches covered with clay. Each dwelling was divided into three sections, with living quarters sandwiched between a cattle stall and a lofty storage barn. Ditches and earthworks protected the settlement from human predators,

The lack of decoration on this Mesopotamian pouring vessel—3700 to 2850 BC—suggests it was destined for everyday use. The introduction of turntables made it possible to mass-produce such complex shapes.

while wattle fences safeguarded crops from foraging animals that roamed the area.

The new climate that had welcomed agriculture into Europe was a fickle host. The plentiful rainfall meant that, unlike in the Middle East, crops would grow almost everywhere without need for irrigation. On the other hand, too wet a season could flood large areas of river valley and low-lying coastland. In a bad year, for example, the Danube River might rise as much as thirteen feet, causing its tributary, the Tisza, to flow backward, inundating farmland for more than fifty miles of its length. It would have been scant comfort to Europe's riverbank dwellers, as they apprehensively watched water levels rise, to know that, elsewhere, the changing weather pattern was having an even greater effect.

In Africa, the end of the Ice Age had had a dramatic impact. The Sahara, which was previously too dry and inhospitable to sustain life, now experienced periods of heavy winter rain blowing in from the Atlantic Ocean. The desert shrank to a narrow strip, surrounded by wooded uplands and grassy plains in which human hunter-gatherers competed with new indigenous fauna such as cattle, giraffes, and elephants. New lakes and rivers appeared, while earlier water levels changed dramatically. In 12,000 BC, Lake Chad had not existed, but within 6,000 years, it had swelled to cover almost 400,000 square miles—an area ten times greater than its present size. At one point, the waters of Lake Turkana in Kenya swelled so dramatically that they overflowed into the Nile watershed, and the Nile River itself rose to such remarkable heights that one spectacular flood deposited sixty-five feet of silt on the Sudanese flood plain of Wadi Halfa.

It was the prodigious flooding of the Nile that was to form the basis of Egyptian agriculture. Set in a narrow, flat-bottomed valley, the northern stretches of the river were surrounded by arid territory. Every September, however, the river would burst its banks, flooding the valley floor before retreating the following month, leaving in its wake a layer of fertile silt. This moist earth was ideally receptive to domesticated wheat and barley, which reached the region from the Fertile Crescent about 5000 BC. And as the population grew, Egypt's farmers soon learned to capture the annual floodwater and divert it to those areas that remained relatively dry. Low, clay-lined stone dikes were constructed to retain the retreating October waters, which were then diverted by sluices and canals throughout the valley. By 3000 BC, the valley was a checkerboard of carefully administered irrigation basins whose waters suc-

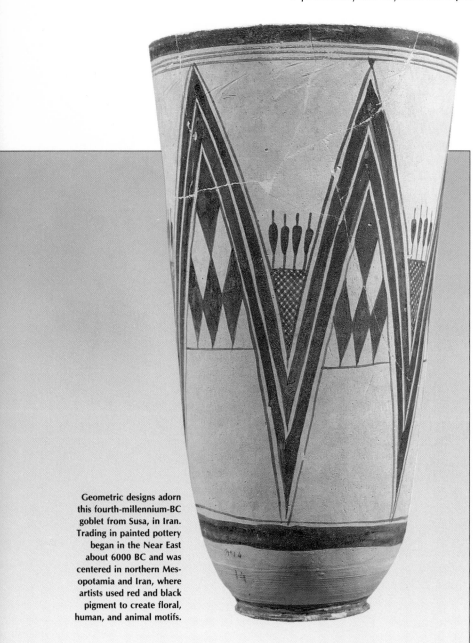

Geometric designs adorn this fourth-millennium-BC goblet from Susa, in Iran. Trading in painted pottery began in the Near East about 6000 BC and was centered in northern Mesopotamia and Iran, where artists used red and black pigment to create floral, human, and animal motifs.

cored one of the most successful civilizations of the time—the Egypt of the pharaohs.

Elsewhere in Africa, agriculture got off to a slower start. Shortly after the arrival of wheat and barley in Egypt, the same crops were being grown by nomads in the eastern part of the Sahara, probably along with sorghum, whose large flower heads ripened to glossy seeds that could either be made into a syrup or ground into flour. Livestock began to be herded, and agriculture gradually established itself in regions as far west as Senegal.

For nearly 2,000 years, these pastoralists of the Sahara settled into a regular pattern of growing at least some of their food and herding cattle, yet still hunting and gathering. They were an artistic people who used catfish spines to decorate their pottery with distinctive wavy lines and decorated exposed rockfaces with paintings that vividly illustrated the animals so important to their lives. But by 2000 BC, the first flush of postglacial warming had receded, and the Sahara reverted to the arid wasteland of today. The lakes dried, the grasslands withered, and the herders retreated, leaving their art to posterity and the windblown sands of the desert.

Those settlers who traveled north would have found a fairly familiar environment in the mountains along the southern coast of the Mediterranean Sea. For those who emigrated southward, however, tropical Africa offered completely different surroundings. Here, the extremes of climate were favorable to a type of vegetation that was a far cry from the vegetation of the Middle East. In Nigeria, for example, the peanut and pigeon pea were probably cultivated to sustain human settlement, as were the plentiful yam and the ensete, or "false banana," whose stem and root provided medicine and fiber as well as food.

The domesticated crops from the north could not gain even a tenuous foothold in the woodland clearings, riverbanks, and forest edges where the native plants thrived; this time, the expected agricultural boom never occurred. Despite the apparent lushness of the region, the newcomers found that the environment was often unsuitable for crop cultivation. Wheat and barley, so important in western Asia, would not grow; they needed winter rain, and in the tropics, the wet season was the summer. In addition, there remains only scant evidence of any indigenous crops being cultivated, although this may be due to the fact that root plants, such as the yam, leave virtually no archaeological remains, whereas traces of seed plants, such as grasses and cereals, often survive either in the form of carbonized grain or as seed-shape impressions in pottery.

Poor soil also hindered the would-be farmers. Since the dawn of time, the warm rains that drenched the tropics had been washing essential nutrients out of the ground. Even the flinty silica—a rocky residue that helps retain nutrients and lends a gray color to soil—had been removed, leaving the earth a barren red. Only in the tropical forests was a suitable reservoir of nutrients retained. But as farmers slashed and burned their way through the trees, they merely cleared the way for further impoverishment. Soon, the red soil would once again be washed out and infertile.

While they were denied the prospect of good yields, the farmers of tropical Africa were also spared the pressure of large populations. Here, disease, particularly sleeping sickness spread by the tsetse fly, prevailed wherever a forest clearing was opened up for cultivation. And whereas irrigation had been so successful in places like Egypt and southern Mesopotamia, the introduction of standing water and irrigation ditches in tropical Africa resulted in a widespread incidence of river blindness and schistosomiasis—a disease, carried by waterborne parasites, that attacks the internal or-

A hollowed-out hearth forms the focal point of this typical twenty-by-twenty-foot Skara Brae house. At the back stands a stone table, on which the family could display their most prized objects—perhaps a necklace made with the teeth of sea creatures, a stone ornament, or the two bone objects below, which might have been used as dice.

THE STONE HOUSES OF SKARA BRAE

The third-millennium-BC inhabitants of Skara Brae, a small agricultural community on Pomona, the largest of the Orkney Islands, off northern Scotland, lived in stone houses built within mounds of solidified refuse. They hewed their homes from a great dome of midden, a mixture of vegetable matter, dung, shells, and stones, which, when left to decay, eventually acquired the texture of hardened clay. Within this dome, they fashioned their living quarters, joined by interconnecting passages and lined by dry-stone walls on both the inside and outside to provide a sturdy, weatherproof environment. No trees grew on the island, so the people of Skara Brae probably used driftwood, supported by whalebone and thatched with turf, to roof their homes. The scarcity of wood also meant they made their furniture out of stone, using bracken for bedding and animal skins for blankets and curtains.

In spite of their somewhat spartan surroundings, these early Orcadians enjoyed a varied diet. On smoky fires fueled by a mixture of dung, whalebone, and heather, they cooked a number of different meats—beef, lamb, pork, venison, and goat—as well as a wide range of seafood and even seabirds, including the gannet, guillemot, and great auk. Fresh water was supplied by nearby streams, and milk came from the islanders' herds of cows and goats, as well as from their flocks of sheep.

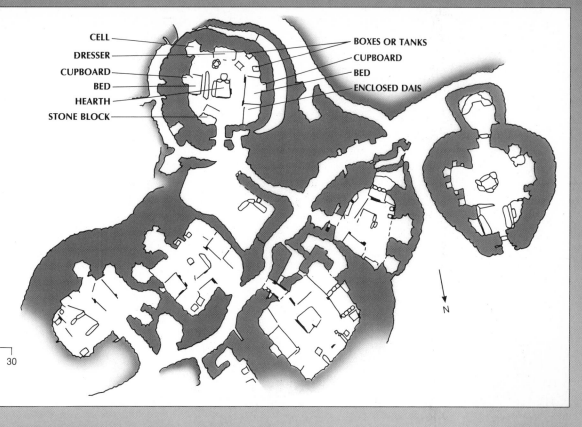

A ground plan outlines the shape of Skara Brae in 2800 to 2500 BC, itemizing the interior features of the house pictured at right and demonstrating the marked similarity in size of the houses. Only one structure, on the far right of the plan, stands apart from the rest; stone fragments littering its floor indicate that it may have served as a workshop.

CELL

DRESSER

CUPBOARD

BED

HEARTH

STONE BLOCK

BOXES OR TANKS

CUPBOARD

BED

ENCLOSED DAIS

N

0   10   20   30

SCALE IN FEET

Passageway with entrance to a house at left

Remains of a bed box, with a storage area set into the wall behind

gans—as well as malaria, all of which had a marked restraint on the numbers of humans and animals living in the area.

Nevertheless, the savannas and forests of sub-Saharan Africa offered huge areas into which farming could expand. With the pressure of disease keeping numbers down, small family groups could split off from the main community and create their own villages. This steady migration to new locations disseminated agriculture throughout the continent. But it was a hard slog. Not until the beginning of the modern era did farmers break through the tsetse belt into the more hospitable climate of southern Africa, and even then, it would be .centuries before the tiny settlements grew into centers comparable to those of the Middle East.

The hunter-gatherers of the Fertile Crescent had not been the only ones to feel the pressure for change. Around the world, unsettled communities were beginning to move in the same direction and, independently, were finding new ways to control the environment about them.

Farming villages first arose in the north of China, thanks—so legend has it—to the endeavors of a mythological leader, Shen Nung. Realizing that his subjects were becoming too numerous to subsist on the meat of birds and animals, Shen Nung—who knew the taste of a hundred different grasses—introduced agriculture, along with pottery making, textiles, and markets. It was a time of peace and prosperity, in which people had no need for defensive walls or weapons.

The myth was remarkably close to the truth. For, by about 6000 BC, farming as a new way of life had sprung up in the north around the middle stretches of the Yellow River. It was fertile ground, free of dense forest, and composed of a rich, yellow loess that retained its fertility well, yet was so fine that it could be tilled with only the simplest digging sticks. The only cereal that flourished in any quantity was millet—a fuzzy-headed grass whose short growing season was ideally suited to the region's climate of cold, dry winters and wet summers.

In this rich land, small settlements grew up, and their inhabitants were able to cultivate two species of millet as well as a variety of fruits and vegetables, including cabbages, plums, and hazelnuts. Hemp provided fiber for clothing, and the cocoons of wild silkworms were plundered for their sturdy filament, which, unlike flax, needed no spinning for extra strength. There were no herds of cattle, but domesticated dogs and pigs provided a supply of meat, which was augmented by wild game from the surrounding countryside.

In Banpo, for example, a village set beside a tributary of the Wei River, 600 people were living in round, conical huts ten to sixteen feet in diameter, some at ground level, others with sunken floors that lay well below ground level. From the doorway, a ramp led down to the earthen floor, which the inhabitants compacted to a hard consistency, probably with the help of wooden mallets. The walls were made of a mixture of mud and straw that was supported by six posts, and a hearth was cut into the floor in the center of the hut. Beside some of the houses, deep storage pits were dug into the ground. The whole settlement was defined by a rough ditch, beyond which lay the village cemetery and a kiln-filled area for pottery making.

At first, life for farmers such as those at Banpo was relatively easy. The region contained only a small population, and there were plenty of new fields nearby to take over when old ones became exhausted. Once again, however, living in a settled community proved to have its drawbacks. Overreliance on millet probably led to dietary deficiencies, and the inhabitants of Banpo became a small-boned people. It was the second millennium BC before the situation was rectified by the introduction of the soybean from Manchuria in the north. Not only did this nutritious bean provide essential oils and protein, but it also absorbed atmospheric nitrogen into the soil, and the people of Banpo soon learned the art of rotating beans with millet to maintain the fertility of the fields.

A second area of early Chinese farming developed around the rich delta of the Yangtze River, near present-day Shanghai. It was here, around 5000 BC, that wild rice, probably introduced from the wetlands of Southeast Asia, first became domesticated in China. The warm, moist climate was ideally suited to the cultivation of this semitropical plant, which reached maximum productivity when its roots were under water. Accordingly, where the Yangtze meandered through the plains toward the sea, the Chinese diverted some of its flow into artificial pools, or paddies. Into these submerged pastures, they would then transplant the individual rice seedlings, which had first been nurtured in dry ground to a height of about twelve inches. When ripe, the whole plant had to be removed from the paddies by teams of stooping workers. The crop was good, and populations burgeoned—which was just as well, for suc-

A Saharan rock-painting from Jabbaren, an area bordering modern-day Libya and Algeria, shows early herders driving two distinct types of cattle: *Bos africanus*, with thin, crescent-shaped horns, and *Bos brachyceros*, with thicker, forward-pointing horns. Dated between 5500 and 2000 BC, when the Sahara was more fertile than it is today, the painting depicts a well-established pastoral lifestyle in which herding and farming have clearly replaced the hunter-gatherer lifestyle that had prevailed in the area since about 9000 BC.

cessful rice growing required the work of many hands—eventually expanding away from the flatlands to cover the hillsides with cultivated terraces.

The drawbacks of agriculture were no less apparent in China than they were in the Middle East. In addition to the normal health hazards of settled life, the Chinese suffered the increased risk of infection from such waterborne parasites as liver flukes, which inhabited the rice paddies. And whatever truth lay in the utopian peace of Shen Nung's time was soon dispelled. More than one excavated village shows signs of its inhabitants being put to violent death by marauders.

The Chinese were not the only ones in Asia to adopt farming independently of the Middle East. Since 7000 BC, the inhabitants of Mehrgarh, a small village on Pakistan's Kachhi plain, had been cultivating wheat and barley and herding livestock. And within three millennia, the settlement had become an important administrative and trading center whose artisans worked in lapis lazuli, marine shells, and turquoise imported from Iran. By this time, too, settlements had sprung up to the east in India's Ganges Valley, where the settlers supplemented a predominantly fish-based diet with their own cultivated rice. It took several thousand years, however, for these advances to reach southern India. Here, the region's inhabitants set up a hut encampment on a valley hillside. They constructed terraces with dry-stone walling, behind which earth was tilled and cultivated. Small, circular huts were built on each terrace— simple structures of wattle and daub—with the village extending from one level to another, upward for hundreds of feet above the valley floor. The villagers grew millet as a main staple crop, which was probably shared as fodder with their large herds of zebus, or humpbacked cattle. Distinct from the wild aurochs of the Middle East and Europe, these beasts were well adapted to a tropical climate, with large surfaces for cooling—a long, narrow head; heavy dewlap; long, pendulous ears; long legs; and a characteristic fatty hump on the back of the neck, which served both as a water reservoir and as a store of energy when food resources were scarce. The domesti-

In a third-millennium-BC stone frieze found in the temple of the birth goddess Ninhursag, in modern-day Iraq, two limestone figures of Sumerian priests sit milking cows *(below, right)* while tethered calves look on, snouts muzzled to prevent them from nursing. In the left-hand section of the frieze, beside a cowshed, fellow priests are probably processing the milk into butter: The first *(far left)* stirs a large jug of milk, which his two colleagues strain into a pot; a fourth priest churns the strained cream into butter.

cation of these animals became widespread throughout the subcontinent, and they were introduced to the Middle East and eventually through Egypt into Africa.

By the end of the second millennium BC, agriculture had blossomed throughout India and Southeast Asia. From China, it had spread eastward to Korea and north into Manchuria. It took a while longer for farming to reach Japan, where the rich diversity of food resources enabled hunter-gatherers to survive quite adequately. Not until the last few centuries BC did population pressure drive the settlers in Japan to cultivate crops—the most notable of which was rice. And there the spread stopped. Many millennia before, Asian hunters had migrated over Beringia to populate the American continent. This time, however, the Americans were going to blaze their own trail.

Across the Pacific Ocean, in Mesoamerica, there were neither the herd animals nor the fields of wild wheat and barley that had encouraged the Middle Eastern people to settle and establish villages. Nevertheless, the region's terrain ran the gamut of growing conditions, from the dry heat of the plain to the cool moistness of the mountains. From the many varieties of edible wild plants that flourished there, the Mesoamericans were able to cultivate their own crops.

For four millennia, Mexicans in the dry, northeastern region of the Sierra de Tamaulipas experimented with plant cultivation, until by about 3000 BC, their efforts in the fields provided one-quarter of their food supply. The earliest plant they harvested was squash, and it was not long before the bean became an important staple. Like the wild cereals of the Middle East that shattered when ripe, the wild bean pod twisted into a spiral and then catapulted its seeds in all directions. As with the wild cereals, there were mutant pods that failed to employ this explosive way of dispersing seed, and these were picked and planted to produce a more easily harvested crop.

Together with potatoes, peanuts, avocados, and chilies, these crops were culti-

vated in small gardens near the settlement. Nevertheless, the transition to full-blown farming was, as in the Middle East, a gradual process. Wild deer and the piglike peccary were still hunted with stone-tipped wooden darts hurled with the aid of a throwing stick; mud turtles were caught; and rabbits were killed in communal drives. When large animals were not found, the hunters' ovens—pits filled with heated stones—were filled with such delicacies as lizards, grubs, and grasshoppers.

It was 1,000 years later and 300 miles away, in the Tehuacán Valley of central Mexico, that the scales first began to tip in favor of farming. As had the settlers at Tamaulipas, the Tehuacáns had steadily experimented with the cultivation of wild plants, in particular a tall grass that thrived around the streams that spread across the valley floor. Initially, it must have seemed an unpromising source of food: The plant yielded tiny ears, about one inch long, bearing just a few husk-covered seeds. But in a relatively short space of time, this scanty offering was to become recognizable as modern-day corn, one of the major building blocks of American agriculture.

In the case of most other plants, it had been evolutionary backward mutations that permitted cultivation to flourish on a large scale. With the Tehuacán grass, however, nature's own efficiency came to the aid of humanity. Unlike the cereal grasses of the Middle East, those in Mexico became cross-pollinated, and as a result, hybrid varieties appeared with longer ears and more numerous seeds. Planted in large numbers by Mexican farmers, these varieties, in turn, released their pollen to produce further hybrids. As food production swelled, so did the population, and the valley inhabitants soon found it worthwhile to make special clearings in the tangle of wild vegetation where crops could be planted on a more extensive scale. By 2000 BC, the seed-head of corn had changed from something a little larger than a thumbnail into a cob two inches long. More important, the transformation had more than tripled yields to 175 pounds of grain per acre. It was enough to turn the tide, and to enable people to settle in villages as farmers.

As the good news of corn cultivation spread throughout Mesoamerica, settlements grew quickly. To their diet, the villagers soon added migratory ducks, the indigenous dewlapped turkey, and small, hairless dogs bred especially for the table. And as in the Middle East, irrigation was used to boost production. In the southern Oaxaca basin, for instance, the water table was in some places just ten feet below ground level. It was an easy matter to sink wells from which water could be drawn and channeled through ditches to reach the fields. Using such methods, two or even three crops could be harvested each year.

Mesoamerica was the seedbed from which farming gradually expanded northward throughout the continent. At the same time, some 1,500 miles to the south of Mexico, the inhabitants of present-day Peru were finding their own way up the farming ladder.

The terrain and climate of Peru were even more varied than that of Mexico. Although located only a few degrees north of the equator, the high valleys of the Andes Mountains provided a cool, fertile environment that teemed with game—deer and two members of the camel family, the guanaco (an ancestor of the llama) and its thick-fleeced cousin, the alpaca—and a number of edible plants, especially the wild potato. In the river valleys that ran down from the hills, there were wild grasses and other food plants, as well as a number of small animals such as the guinea pig. On the Pacific coastline, however, the country consisted of semidesert that received appreciable amounts of rain only one year in every twenty-five.

The high Andes were the heartland of South American agriculture. It was the early hill farmers who first began to cultivate the potato, learning how to preserve the tuber for storage. In the cold mountain air, the potatoes were left to freeze and thaw alternately, while being trampled underfoot to drive out all moisture. The result was a fully dehydrated food that could be stored or ground up into flour. These were the same people who, in 3500 BC, saw the advantages of domesticating the guanaco. These sturdy, all-purpose beasts provided wool for cloth, meat that was either eaten fresh or sun dried, sinews that were used for binding, and dung that served as fuel and fertilizer. But above all, the animal was capable of carrying loads of up to 130 pounds more than ten miles a day. Within 2,000 years, the alpaca had joined the llama as a mainstay in a herding economy that stretched along the western seaboard of Peru.

All along the Pacific coastline of Ecuador and Peru, farming settlements were established. Where soil showed signs of impoverishment, people restored its fertility with guano—the accumulated excrement of the millions of seabirds that inhabited the rocky coastal islets. And once again, irrigation began to work its magic. It started at the coastal deltas, where the inhabitants dug simple ditches down which they diverted the Andean rivers' annual floods away from the sea and onto their fields. As the population grew, so the ditches snaked farther and farther toward the river source until, eventually, they reached the mountains. By about 1000 BC, the crude ditches had become an extensive canal network, winding around the contours of the hillsides for about seventy-five miles. Fifty feet above the ground, hand-built aqueducts of stone and earth, more than one mile long, spanned entire ravines, bringing much-needed water to adjacent valleys. These edifices meant more than just an increased food supply; they represented the first stirrings of civilization in South America.

A sandstone engraving *(right)* found in the French Alps depicts a farmer of the third millennium BC using two yoked oxen to pull a rudimentary plow; the stylized horns of the oxen are clearly visible in the outline drawing below. The use of animals for hauling came only after human beings had already learned to exploit animal skins for clothes, their bones for tools and building materials, and their meat and milk for sustenance.

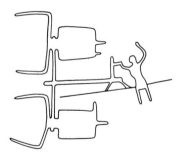

As was the case in the Old World, so in the New, the adoption of agriculture brought with it a new range of health problems. At the Dickson Mounds, in Illinois, there is evidence that people became increasingly dependent on corn as the principal foodstuff. The skeletons of these people illustrate the hazards of such a policy: marked iron deficiency and an upsurge in dental defects caused by the lack of a properly balanced diet.

The prehistoric communities of the Georgia coast tell a similar story. Here, the early farmers' remains show that the health of women, in particular, declined with the introduction of agriculture. Whereas the female members of the community stayed at home tending the corn and consumed a limited diet, the males still went out on occasional hunting and fishing expeditions, on which they had the opportunity to eat protein-rich meat and fish.

Agriculture had come a long way in the 6,000 years since the mutant grains had first been noticed growing on the garbage dump. It was still far from being a panacea for the world's food problems: Increased clearing of forests, overgrazing, and the use of the plow eventually led to soil erosion. In the Middle East, the irrigation systems that had once been such a boon eventually rendered much of the land useless for agriculture; the fierce heat of the sun had caused the floodwaters to evaporate, leaving behind deposits of salt in such quantities that plants and vegetables were unable to survive. Nor did the advent of a new way of life mean the death of the old. Until its extinction in AD 1671, the aurochs continued to be hunted in the primeval forests of northern Poland, and as recently as the late nineteenth century, the situation in prehistoric Abu Hureyra was paralleled by that in southern Africa, where hunters culled migrating antelopes whose herds, estimated at 10 million, stretched from horizon to horizon. Even today, in the frozen Arctic, in the depths of South America's jungles, and in the deserts of Australia and Africa, communities still pursue the hunting and gathering lifestyle of their ancestors.

Still, once established in farming settlements, for most people, there was no turning back. The pressure of population, which had been one reason why humans had taken to farming, was now an overwhelming obstacle to its rejection. The only way to go was forward. Humanity had learned to control nature; it was now faced with a new challenge: that of controlling itself.

For agriculture had done more than provide an alternative means of survival: It had dramatically changed the fabric of human society. With enough to eat and more time on their hands, settlers could now turn their attention to matters other than food production. Those who had the ability could earn a living from selling the products they made: tools, weapons, burial items, cooking utensils, and personal adornments. As trade boomed, so, increasingly, people lived off one another rather than off the land. And, inevitably, some—those who were better artisans than others, or whose lands were more productive—grew wealthier than their companions. It was a wealth that not only gave them the status of employers but also could be passed down through the family. The beginnings of a hereditary class system had emerged.

More than wealth or opportunity, the new social hierarchy meant authority. Someone had to oversee the construction and maintenance of irrigation systems; to coordinate the settlement's defensive arrangements; to administer religion; and to regulate and record trade. These new titles would emerge as the distinctive hallmarks of society's progression toward urban life.

A NEW DESIGN FOR LIVING

More than 8,000 years ago, a bustling, thriving, trading center known to us as Çatal Hüyük, or the "forked mound," stood overlooking the sweeping plain of what is now south-central Turkey. A copy of a wall painting (*above*) found in one of the town's shrines depicts Çatal Hüyük as it looked around 6700 BC. Closely packed rectangular houses rise in graded terraces, while behind them, the twin-coned volcano of Hasan Dağ, eighty-five miles to the east, belches out fire, smoke, and molten lava.

This busy community of several thousand souls grew wheat, barley, and peas in the surrounding fields. From the hillsides came almonds, pistachios, and fruit, and for their meat, the townspeople not only hunted the herds of wild pigs, red deer, and aurochs that roamed the open flatlands but also ate

the cattle and sheep that they had domesticated. The town's prosperity may have stemmed in part from the region's plentiful obsidian—a shiny, black volcanic glass used to make knives, mirrors, and jewelry—which was excavated at Hasan Dağ and traded for items such as decorative Mediterranean seashells and good quality Syrian flint. Çatal Hüyük may also have been a religious center. Only one of the town's thirty-two acres has been uncovered, and in this area alone, one building in three seems to have been a shrine.

Their apparent piety notwithstanding, many of the town's inhabitants evidently failed to ward off misfortune. Their remains show signs of the skull-thickening caused by malaria, of injuries sustained in rooftop falls, and of head wounds caused by blunt instruments. For the most part, though, the populace of this fledgling urban community enjoyed a level of security and sophistication unattainable elsewhere at the time, as depicted on the following pages in illustrations that are based on archaeological finds made at the site.

A TOWN WITHOUT STREETS

The people of Çatal Hüyük built their houses so close together that the rooftops had to serve as the town's thoroughfares, with wooden ladders providing a means of getting from one level to another *(right)*. Entry to each house could be obtained only through a small door on the second story. With no access at ground level, the town presented a solid bulwark of blind walls to both floods and attackers—a feature that made Çatal Hüyük a far safer place to live than most settlements at that time. When it rained, gullies between the houses drained the water off the roofs; down below, the sites of derelict houses served as garbage dumps and latrines, over which the townspeople sprinkled layers of ash to curb the odor. Although the buildings all looked somewhat uniform, each household had its own distinctive clay seal *(below)*, for marking clothing or bags of grain—or even for decorating its owners' skins.

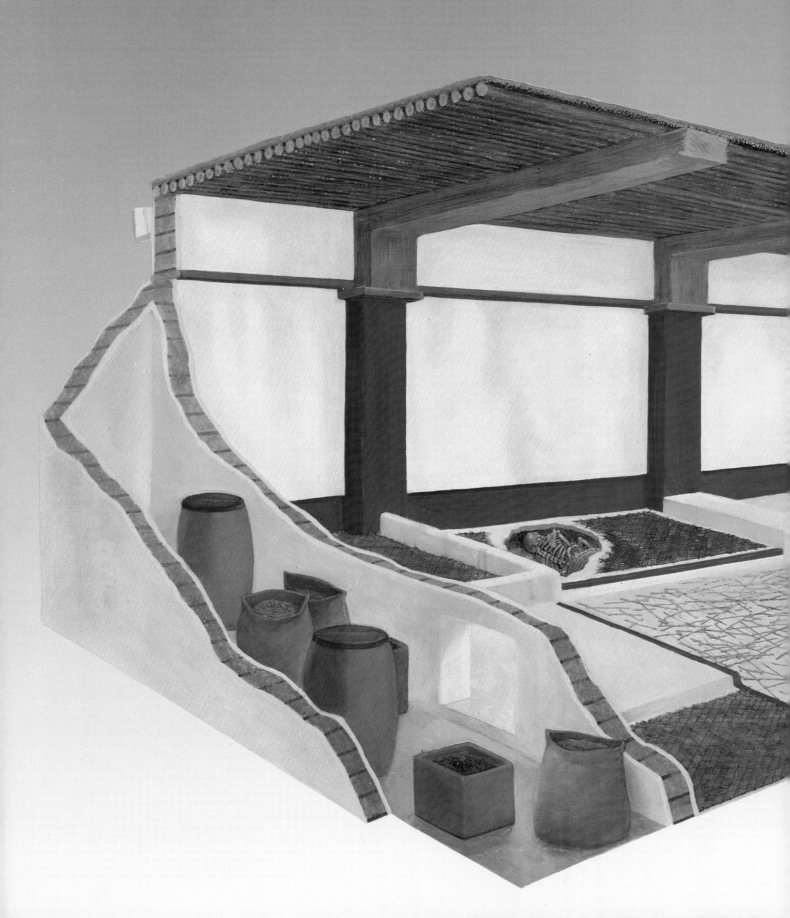

A cutaway reveals a skeleton *(middle ground, left)* wrapped in a basket and buried under the spot where the mother and children slept. Mourners first exposed a corpse on a platform outside town, where vultures picked the bones clean. Then the family took the skeleton home for burial, perhaps during a spring festival in which townspeople repaired their houses.

This is a typical family home—one room about fifteen by twenty feet. A rough-hewed staircase hung with pig bladders containing drinking water descends from the roof. Two oak rafters support crossbeams covered by mats of woven grass, bundles of reeds, and an outer layer of dried mud. A door in the inner wall leads to a storeroom housing clay chests and jars. Under the stairs is the kitchen, with an open hearth, an oven set in the wall, and a variety of utensils, such as a burnished pot, a bone spatula, and a two-pronged fork *(below)*.

A DWELLING FOR THE DEITY

Symbol of male potency, the plaster head of a bull, flanked by rows of plaster female breasts, thrusts out of the wall of this Çatal Hüyük shrine. A painting on the left-hand wall depicts giant vultures with outspread wings picking a headless corpse clean prior to burial; on the floor, skulls of the dead sit in mute homage to the shrine's deity. Devotional objects excavated at Çatal Hüyük include a ceremonial flint knife, with its bone handle carved in the form of a snake, and a marble statuette of what is thought to represent a mother-goddess (below).

ARCHITECTS OF A NEW SOCIETY

Without the water from their spring to refresh them, the people of Jericho could never have found the strength to build the wall that encircled their town in 8000 BC. The digging of a broad defensive ditch and the gathering and piling up of stones into a rampart was thirsty, painful work. The construction of a massive watchtower, almost thirty feet high, would be even more arduous; there would be angry words, accidents, perhaps even deaths. But these new defenses were a necessity. Their settlement, near the northern end of the Dead Sea, was an oasis in the arid land that later generations would call Palestine. Their successors would wage war over ownership of the region and its water sources for another 10,000 years or more.

To nomadic bands emerging from the southern deserts or to migrants descending from the northern heights, the huddled oasis settlement of round, mud-brick houses exuded an enviable prosperity. In 8000 BC, Jericho's fields, nourished by the stream that flowed from the ever-generous spring, grew emmer wheat in abundance; the surrounding wilderness, with its population of gazelles and ibexes, provided a seemingly unending supply of fresh meat. As many as 1,000 people—an astonishing prospect in a landscape still populated mainly by small, struggling settlements or wandering tribes—lived here together.

But Jericho's Neolithic inhabitants could do more than simply feed themselves: They possessed a surplus of material goods that they could send away in bags or baskets on the backs of traders, to be bartered either for materials they could not find on their own home ground, or for some tool or utensil crafted in a manner with which they were not familiar. Prosperity gave them the opportunity to stand back and consider their shared situation; they were organized enough, and articulate enough, to cooperate in enterprises for the common good.

Faced with an expanding population, the people of Jericho had agreed upon a method of channeling the stream that ran below the settlement in order to water the fields and increase the yield of the wheat crop. One successful engineering project quite possibly engendered another. They had learned how to pool ideas and effort, how to make a plan and achieve results, as well as how to construct a wall to protect themselves against attack.

In the end, though, their ramparts did not save them. For generations, perhaps for many centuries, the people of Jericho kept their defenses in good repair, but at some point during the eighth millennium BC, their community simply disappeared. The cause may have been either some natural disaster or an invading tribe that swept the inhabitants away and left no trace of its attack. Yet, a site with so many natural advantages would not remain empty forever. Within a few centuries, possibly even sooner, a new population, with no apparent links of kinship or culture to its prede-

An inhabitant of Jericho, a city northwest of the Dead Sea that flourished around 10,000 years ago, is immortalized in this lifelike head, modeled in plaster over a human skull. A number of such heads have been found buried beneath house floors; with cowrie shells for eyes and realistic facial features, they appear to represent particular individuals. The heads may have been intended simply as family mementos, or they may have played a part in a cult of ancestor worship.

The world's first settled societies, which came into being between 10,000 and 2500 BC, were concentrated in four main areas: the Near East, the Nile Valley, the Indus Valley, and the North China plain. Sited on or near rivers within a belt of largely equable climate, between latitudes twenty-four and thirty-eight degrees north, these clusters of towns and cities owed their origin and survival to abundant agricultural supply from the surrounding region.

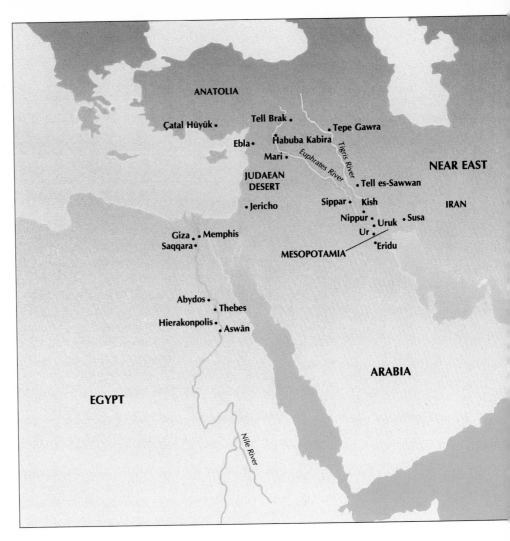

cessors, established a new settlement on the mound that now concealed the ruins of the old community.

For thousands of years, Jericho's spring would nourish the crops of a succession of different populations, who were in the process of either moving down from the mountains or else emerging from the desert. Some would be better farmers than their predecessors, others less advanced. Some would be skilled at building solid houses, others—centuries after them—would camp in roughly dug pits, roofed only with animal skins stretched on frames. Some would be gifted potters, whose bowls and drinking cups were as beautiful as they were functional; their successors might rely only on the crudest, most primitive ware. But each group would learn new skills in its own time and build upon that knowledge. Then some disaster would occur—a plague, an earthquake, or a shrieking horde of raiders—and these people, too, would be gone. Sand would cover their ruined houses, mud would clog the irrigation channel, and the ancient mound would stand empty until new arrivals came to build on it once again.

This pattern would repeat itself, not only in the neighborhood of Jericho, located at a crossroads of many migratory routes, but on innumerable other sites in every

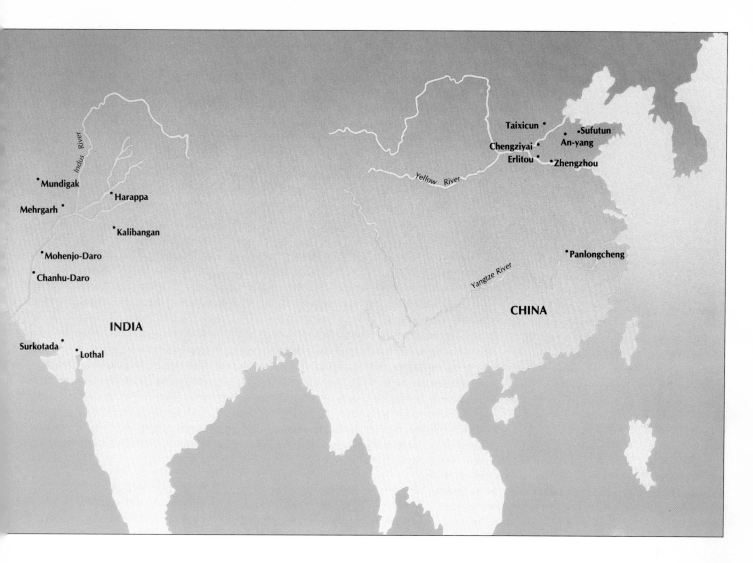

inhabited land. In the Judaean desert, as in China, the Indus Valley, and the Nile Delta, civilization was not destined for an easy birth. At times, it seemed that every step forward was followed by two steps backward. Humankind, waking to its possibilities, did not march along a single, straight road but threaded its way through a succession of mazes composed of switchbacks, dead ends, and endlessly proliferating branches. But still it progressed.

The first wall builders of Jericho, hoisting their stones into position, had been merely a single cell in the revolution taking place throughout the inhabited world between the tenth and third millennia BC. During this era, the people of the earth began to reap the rewards of their own intellectual and technological advances. They began to take control of their environments, producing their own food and creating new tools to ease and speed their labor. They discovered how to live and work together in large communities, and they began to trade—not only goods, but also knowledge—between one settlement and another. At the beginning of this timespan, human life was lived beneath the constant shadow of famine and disease. By its end, communities were still just as likely to be wiped out by starvation or sickness, or by the new scourge of warfare, but culturally and socially, they had moved far beyond their predecessors. Civilization had begun.

The engine that propelled these movements was the community. As women and men learned more about the world, the words and phrases they used to exchange ideas and information grew correspondingly more complex. The past could be remembered, the future planned. People who were descended from the same tribal ancestors, or else had settled close together, understood each other; they saw the world, given shape by their words, in a similar way. Those from another stock, or those who migrated far afield and kept no contact with their kinfolk, employed different sounds with different meanings. Even those tongues that evolved from a common root soon diverged until all mutual comprehension was lost. The paths that these ancestral languages traveled were as varied as the paths their speakers took to populate the earth, and as subject to speculation. But words, at least those uttered in the time before writing was invented, just disappeared into the wind; unlike stone shelters, flint tools, bone ornaments, and even the action of fire upon wood, they left behind no traces.

Nevertheless, something did survive to indicate to later generations the ways in which these early humans thought and acted. Their sites of worship remained. Massive boulders were dragged many days' journey from the places where they had been quarried, to be either set up in circles or aligned in rows that strode across the landscapes of the British Isles and Brittany. Other, later edifices, carved with the somber faces of the gods, stood vigil on remote Pacific islands, or vanished under the green prolificacy of a Central American jungle. In eastern Ireland, a richly decorated burial chamber was carefully positioned to receive within its depths a single, hopeful shaft of sunlight on the darkest day of winter. In the British Isles, stone slabs, twice

Occurring naturally in lumps and seams as well as in ore, copper was the first metal to be put to use by humans. As early as 9500 BC, smiths in southeast Europe and western Asia were hammering the metal cold to form ornaments, tools, and weapons. Later, they discovered that heating the metal made it stronger and more malleable, and around 6000 BC, they learned to extract copper from ore by applying intense heat. Still greater temperatures were required to melt the copper, so that it could be poured into clay or stone molds to cast tools such as knives and axes. Further advances enabled smiths to work with bronze—a mixture of copper and tin—iron, and by 150 BC, steel. Examples of their increasingly sophisticated work appear on these pages.

WORKING WITH METAL

A hammered copper blade from Syria is flanked by two delicately wrought, cast toggles, one with the head of an ibex or gazelle, which were probably used for fastening cloaks.

In a frieze from Tell el-Ubaid in Mesopotamia, a lion-headed eagle representing the storm-god grips the tails of two stags. Dating from about 2700 BC, this carved-wood panel is covered with beaten copper, secured by copper nails. The stags' copper antlers are soldered in place with lead.

as tall as any person, were planted in pits in the ground; two, three, and sometimes as many as sixty of them were spaced at regular intervals. The people who had set them in place chose configurations that might have served as a massive sundial; other arrangements could have been employed to chart the slow, seasonally changing procession of stars across the night sky.

Whatever their function—the honoring of unseen powers, the measurement of time, the commemoration of some great event—these relics spoke of shared beliefs and common purposes. So, too, did the images that people painted on rocks, which they carved and kept in niches in their houses and tucked lovingly into the burial pits of their dead. The deities they worshiped varied in form and gender. Some were given the shape of beasts; others were faceless entities, too potent to contemplate directly and represented only by symbols.

The newcomers to seventh-millennium Jericho arrived on the mound many years after the disappearance of their predecessors. The work of wind and weather, soil erosion had covered the traces of this earlier community so completely that its successors could barely make out the remains of earlier walls and houses. These new people brought with them a uniform style of architecture and a conformity in their domestic arrangements that suggested that they were not nomads, but veterans of a settled life displaced from some unknown homeland. They built their large rectangular houses from hand-molded, sun-dried mud bricks that were the shape of flattened loaves; these were held together by mortar, at that time a new invention, which was probably made by mixing heated limestone with sand and water. On the upper surface of the bricks, the Jerichoan brickmakers would impress a row of thumbprints in a herringbone pattern; when construction began, the builders would pour the

mortar into these indentations until it began to overflow, at which point the next brick would be placed on top.

As prudent housekeepers, the Jerichoans covered the clay floors of their houses with tough, hardwearing lime plaster, which was colored cream or red. Those who were responsible for the housework evidently had a say in the architectural design too; the plaster and the wall met not at right angles, where dust and dirt would collect in the crack, but in an upward curve that would allow dust no place to hide from broom or leather mop. Outside, these diligent housekeepers placed large vats against their walls to collect precious rainwater and did their cooking in a central courtyard, grinding einkorn wheat in querns, and preparing plenty of meat. Like the very first occupants of Jericho, they relished the gazelles brought in by the community's hunters, but they also ate the flesh of sheep and goats that were either wild or in the earliest stages of domestication.

Naturally, the fertility of their fields, and of the beasts they either reared or hunted, were matters of profound importance. In the hope of cajoling, or controlling, the forces of nature, they shaped small animals out of clay and formed female figurines with generous curves—images of fecundity and the mysteries of the creation of life. Certain buildings within their settlement seemed to be set apart for ceremonial purposes; one of these structures had two rounded alcoves in which sacred objects might have been lodged and a central basin that may have received the blood of sacrifice.

The dead were accorded their own kind of deference. Their bodies were taken away from their houses and set apart until the flesh was gone—decayed or stripped away by animal predators.

A Mesopotamian charioteer drives a team of four onagri. This statuette from Tell Agrab, made about 2700 BC, is the earliest-known example of the lost-wax process: A clay mold was formed around a wax model; the wax was heated and drained off; and molten copper was poured into the mold.

Then, the skulls were carefully brought home again and buried under the plaster floor, allowing the house's living occupants to pursue their daily tasks in comforting proximity to their departed kin.

These practices were not unique to Jericho. Skulls found buried in clusters under similarly decorated rooms in smaller settlements in Syria suggest that the Jerichoans may have been members of some larger grouping, allied not only in their spiritual beliefs but also in the language they spoke, the tools they used, and the customs by which they lived.

Increasingly, as populations grew and settlements spread, their members established peaceable contacts for the purposes of trade. Surplus foodstuffs were only a small part of what they traded. Artisans everywhere were ready to pay good prices for the raw materials they needed and the substances that would give color to their finished wares. Beidha, for instance, a small village in Jordan, did a thriving trade in its supplies of the red iron ore hematite, which gave pottery a handsome red color; it could also be used to make the rouge that women rubbed on their cheeks or applied to their lips with tiny spatulas of carved bone. Small settlements in Iran traded lapis lazuli and carnelian, used for jewelry, and soapstone, which was easily carved into smooth, pleasing shapes.

Jericho, with easy access to the Dead Sea's salt deposits just seven and one-half miles away, found itself endowed with a highly prized commodity: People craved salt in their diet and discovered how efficiently it preserved their surplus foodstuffs from decay. Relayed along a chain of agents that might extend for hundreds and even thousands of miles, these goods were exchanged for cowrie shells gathered by dwellers on the Mediterranean seashore; for turquoise from the Sinai Peninsula; or for the hard, glassy black substance known as obsidian, sharp enough for knife blades and shiny enough for mirrors, which came from the slopes of extinct or dormant volcanoes. In the eyes of Jerichoans, obsidian was invested with an added layer of desirability due to its having originated on the distant, mysterious plateau of central Anatolia, 500 miles to the northwest.

A large part of the Anatolian obsidian trade was conducted by the residents of Çatal Hüyük. Unlike Jericho, the site at Çatal Hüyük suffered no disruptive cycles of abandonment, destruction, repopulation, and rebuilding. A single people occupied the site, with no signs of disturbance, for at least 1,500 years; when they did depart, it was only to construct themselves a new town on a site across the river, where they prospered for another 800 years before disappearing from the scene—with no evidence of violence or conquest.

Enjoying the benefits of continuity, prosperity, and tranquillity, the inhabitants of Çatal Hüyük were free to develop an extraordinary repertoire of crafts and talents. Their town was a wonder for its age—an apparently affluent community of scrupulously clean, well-furnished households and elaborately decorated shrines.

Religion, or its outward trappings, formed the backdrop to Çatal Hüyük's domestic life. In at least one part of the town—indeed, by some reckonings this well-ordered and apparently populous settlement would qualify as an embryonic city—one out of every three buildings served as some kind of ceremonial chamber or shrine. Mother goddesses, vibrantly painted or sculpted in dramatic relief, gave birth to divine daughters while resting on the shoulders of sacred leopards; quintessential images of gender—masculine horned beasts and women's breasts—projected from walls or benches. During the settlement's long life, these images demonstrated an apparent shift in values: The horned skulls of male animals, sacred to the hunter, gradually vanished from the holy chambers, first outnumbered and then replaced by the life-giving female, whose fertility mirrored that of the earth itself. The hunter surrendered his primacy to the farmer tending the fields.

In order to glorify their gods, furnish the burial places of their dead, and enhance their own domestic lives, the people of Çatal Hüyük sought far and wide for the materials that their skilled workers needed. Timber for building—oak and juniper—came from distant wooded slopes, where it was cut into logs and floated down the rivers to the plain; limestone, alabaster, and fine white marble were carried from quarries that were many days' journey from the settlement; shells of sea creatures—cowrie, dentalium, whelk—were transported all the way from the Mediterranean seacoasts. Traders, climbing the foothills and negotiating the mountain passes, kept a sharp eye out for signs of the bright minerals and metal ores that painters needed to make their pigments: blue-green malachite, sky-colored azurite, brown and yellow ochers, deep red cinnabar.

The work of its artisans testified to the fact that Çatal Hüyük was a town of specialists; no footsore hunter or hard-pressed subsistence farmer could have spared the time, the energy, or the daylight required to achieve such levels of skill. Every kind of stone was ground, polished, pierced, or carved to make heavy-duty sickle blades and axes, mortars and pestles, adzes or chisels for fine woodwork, palettes for paints and cosmetics, and snake-handled ceremonial daggers that were realistically adorned with serpent's eyes and stippled scales. Carpenters knew how to construct squared roof-beams, doorjambs, and sturdy ladders, as well as wooden tableware of advanced design: large meat platters with carved handles, sauceboats, and tiny eggcups. Weavers used fine rushes and marsh grass to make floorcovers, understood how to turn plants into dyestuffs, and used clay stamps incised with different geometric designs to print color onto cloth. The vivid hues and patterns they worked into their textiles were copied by painters onto the walls of shrines and houses; the same types of design would be preserved and reproduced by the women of the region for 8,000 years to come.

The craftsmen of Çatal Hüyük also possessed another valuable skill—the art of metallurgy. They knew how to apply heat to lead and copper, and how to hammer them into shape, turning out a variety of beads and ornamental cylinders, which might have been strung on pendants or threaded onto the fringes of clothing. They were early practitioners of the craft, but not necessarily its pioneers. The discovery came first, sometime between 7000 and 5000 BC, to settlements scattered throughout western Asia and southeastern Europe; it would be another 2,000 years or more before it penetrated to northern Africa and the Far East.

The knowledge that heat could be used to draw metals from particular rocks, and that these metals could be manipulated into different shapes, moved gradually along the ever-expanding trade routes. The merchants who carried off a bundle of copper beads to exchange for other desirable objects might travel no more than thirty miles from their homes, but the goods themselves could be handed on from one trader to another, until they ended up halfway across a continent. Inevitably, information about their manufacture traveled with them. Itinerant metalworkers, brave and enterprising enough to leave their own people, might speed up the transmission of their skills—or might learn some new technique themselves and broadcast it farther afield. And, less peaceably, invaders who possessed the technology introduced it to the territories they conquered.

Naturally occurring nuggets of copper ore could be picked up from the ground in areas such as Iran and Turkey, where they were to be found in abundance. But the working of these materials would proceed only when artisans began to understand the mysteries of applying and controlling intense heat. The metalworker and the potter were faced with a similar set of technical problems: the construction of ovens to withstand powerful temperatures and the need for molds and tools that could be used without harm to the artifacts or injury to their makers. Kiln-fired pottery predated metalworking, but it is not inconceivable that both crafts took their lead from the art of baking, which used a primitive oven to turn ground grain into bread.

Gradually, the metallurgists discovered more efficient ways of working. From merely hammering out copper nuggets into random, brittle shapes, they learned how to make the material more malleable through the application of heat, at first by placing it in an open fire. Then, as nuggets of copper became harder to find, they also learned to extract the metal from more readily available ores, such as malachite and azurite, by smelting them in an enclosed oven.

It would take another thousand years before they devised the more complicated process of casting copper in open molds. This technique required the development of an oven capable of reaching the intense temperature of almost 2,000° F., the point at which copper itself—and not just its impurities—would liquefy. The molten metal was then transferred into carefully shaped molds of clay or stone and allowed to cool into the desired form before it was hammered into a finished product. Gradually, the techniques for making molds grew more elaborate; artisans learned, for instance, that a charcoal cone could be inserted into a mold to produce a hole for a shaft or a blade handle. By the fourth millennium BC, metalworkers had mastered their art to such an extent that they could manufacture artifacts such as the exquisite little specimen found in a tomb at Tepe Gawra, in what today is northern Iraq: several pieces of electrum—a naturally occurring alloy of gold and silver—expertly welded together to form a smooth, silver-colored wolf's head, one inch long, complete with upright ears and curved, bared fangs.

As more and more people took up this new skill, metals became an increasingly valuable commodity along the trade routes. The nuggets of natural ores that lay on the ground would not suffice to meet demand, and men learned to dig into the earth in the areas where these substances were found. Mines near modern-day Eilat, for instance, at the northernmost tip of the Gulf of Aqaba, consisted of shallow shafts that led to a network of narrow subterranean galleries, dug to follow the course of the ore-bearing strata embedded in the rock. The miners who worked them must have been either very young children or members of some unusually small-boned tribe: The galleries would have been impassable even for the most slightly built fully grown adult of that time.

For those with the resources on hand, or with the wealth to acquire them through trade, metal tools and weapons now became a realistic possibility. Yet, many generations would come and go before metalworkers perfected the technique of alloying copper with tin to make the far tougher, more durable metal known as bronze. This new and much sought-after material would emerge first in the Near East, some time between 4000 and 3000 BC; it would take another thousand years or so to spread to

Polished to a lustrous shine, beads of precious stone are strung in colorful sequence. The upper necklace, from Ur in southern Mesopotamia, includes carnelian, chalcedony, and lapis lazuli. In the lower one, dating from around 5500 BC and found at Tell Arpachiya, in northern Mesopotamia, lozenges of black obsidian alternate with cowrie shells.

the farthest reaches of Europe and Asia. When it did so, it would change the face both of warfare and of society, setting apart the powerful elites, with their bronze weapons and armor, from the common mass of people, armed only with bows and arrows, maces, or axes made of stone.

Meanwhile, as the pioneers of metalwork stoked their ovens and designed their molds, revolutionary social developments were taking place in western Asia, in the fertile alluvial plain that lay between the Tigris and Euphrates rivers.

The society in Sumer known as the Ubaid—which flourished between the fifth and fourth millennia BC in what today is southern Iraq—was blessed with as many handicaps as it was advantages. The soil in the region was rich and easy to cultivate, but the climate was arid, with long, hot, rainless summers. Its southern reaches, where the rivers emptied into the Persian Gulf, were a zone of lagoons and marshes that harbored an abundance of fish and game, hospitable to the useful date palm, but salty and subject to tidal floods. Farther upstream, farmers were confronted with a different problem—how to bring water to their parched fields. Their solution was to

enlist the cooperation of every single farmer who cultivated land along the banks of the rivers and their tributary streams.

Working in concert, the farmers planned and constructed an elaborate system of irrigation channels and organized themselves to ensure that the network was properly maintained. The benefits these farmers gained went far beyond the carefully orchestrated flooding of their ditches: Their successful collective efforts laid the groundwork for other forms of collaboration. These might include either material enterprises, such as the construction of defensive walls and communal granaries, or more spiritual endeavors, such as the building of a great temple to pay homage to the gods who controlled their harvests.

As communities grew and prospered, new, grander types of architecture appeared. As early as the sixth millennium BC, there were houses with fourteen different rooms at Tell es-Sawwan, an evidently affluent village in central Mesopotamia, bordered by the Tigris River and protected both by a buttressed wall and by a ten-foot-deep defensive ditch. In this settlement, as in Çatal Hüyük, people buried their dead beneath the floors of their houses, and the funerary items placed with them display a high standard of craftsmanship in materials that had been expensively imported from distant places: figurines of alabaster and beads of turquoise and carnelian, probably transported from Iran.

At the same time, a striking new type of luxury pottery made its first appearance in northern Mesopotamia. Named after the settlement of Tell Halaf in northern Syria, the style was characterized by bold, linear patterns, checks and zigzags mixed with floral and fish-scale motifs, and the use of a wide range of reds, browns, and blacks in the color scheme. The widespread distribution of the pottery—it has been found 150 miles away on the Turkish Mediterranean coast—shows that, although fragile, it fetched a high enough price to make long-distance trading worthwhile.

The people of Ubaid Sumer lacked many natural resources. They had plenty of reeds and mud, but they had little stone and none of the minerals needed for metalwork. Geography was in their favor, however; people and merchandise could move easily across their flat landscape. Trade routes crossed the increasingly well-settled Mesopotamian plain, linking the Zagros Mountains on its eastern edge with the villages that hugged the fringes of the Arabian Desert on the west, or traversing the wilderness to reach the Mediterranean Sea. Impelled by necessity, the people of Ubaid Sumer looked to commerce as another form of farming, scattering merchants like seed. The pottery vessels their traders used, whether to carry goods for sale or to hold their own provisions, could be found as far apart as the northern reaches of present-day Iraq and the eastern shores of the Arabian Peninsula.

The movement of trade goods between far-flung mines and workshops, farms and fisheries led to improvements in the means of transportation. The domestication of beasts of burden, such as the donkey or the ox, allowed a trader to carry more wares than he could hoist on his own shoulders and to take them farther. It was a short step from the use of pack animals to the invention of the cart. It may have been potters who, in the fourth millennium BC, first employed a wheel to speed the working of their clay; if so, it was not long before someone thought to turn the circle on its side to provide a means of locomotion.

The people who lived along the lower reaches of the Tigris and the Euphrates were certainly seafarers. Simple rafts had been used for thousands of years, but ships with

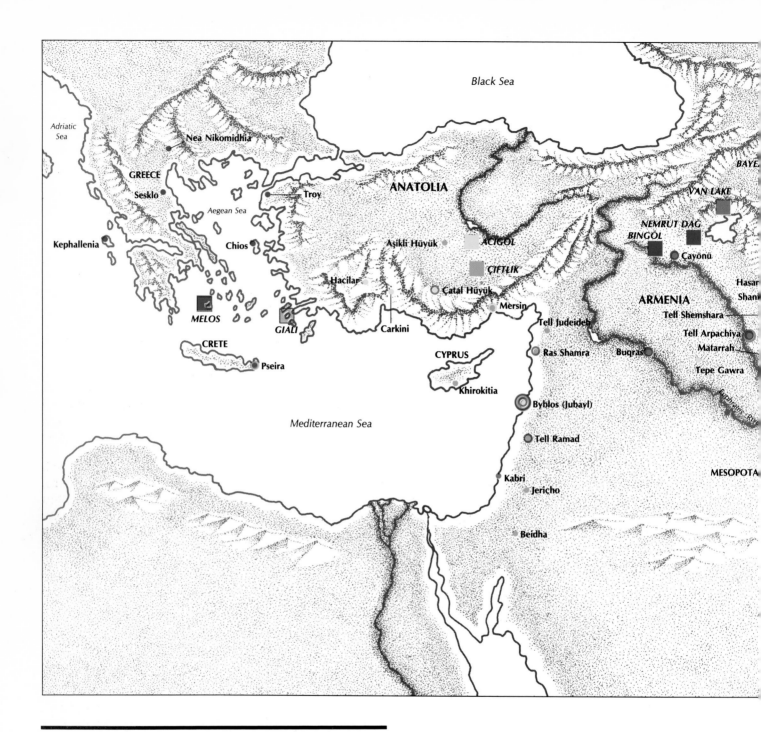

In the period between 7000 and 5000 BC, towns and settlements across the Middle East were linked by trade routes along which merchants transported the prized commodity of obsidian, a hard black volcanic glass formed by the rapid solidification of molten lava and used to make knives and arrowheads before the introduction of metal. The presence in these artifacts of varying amounts of elements such as iron and sodium has enabled archaeologists to pinpoint the source of the stone. The map above shows the location of obsidian-producing areas *(squares)* and indicates which nearby settlements *(circles)* used each source's output. The obsidian from two sources in eastern Turkey—Bingöl and Nemrut Dağ—is indistinguishable, so both are in the same color.

sharp prows and sails first appeared in Mesopotamia during the Ubaid period. After negotiating the lagoons and marshes at the mouths of the rivers, it would have been easy enough for mariners to hug the Arabian coast as they followed a course down the Persian Gulf. Ubaid adventurers undertook these voyages either in search of pearls or shells, or as straightforward commercial fishing expeditions; the traces of their Arabian campsites suggest they visited the same sites at regular intervals, and they stayed ashore long enough to dry and preserve their catch for eventual sale in landlocked markets.

Dried fish, pearls, and decorative shells would all find ready purchasers in Eridu, which the Sumerians would later call "the oldest town in the world." Built on the extreme southwestern edge of the Mesopotamian alluvial plain, Eridu stood on an inhospitable site, constantly menaced by encroaching sand dunes. Nevertheless, it housed a thriving community of specialists, among them not only potters, weavers, bakers, and brewers but those much sought after experts in the all-important mystery of how to please the gods: the priests and priestesses.

For Eridu was a pious place, home of the water god Enki, who was believed to preside over the canals—fed from the Euphrates twelve miles away—that were essential to the city's survival. At the heart of Eridu stood a temple, which would survive for centuries, carefully preserved within the foundations of its larger successors. In these sacred precincts, animals were sacrificed, their blood poured over the altar as a libation. Worshipers brought offerings of food and drink; Enki, it seemed, enjoyed the same refreshments as his mortal subjects—bread, beef, lamb, and beer. The contributions also included seaperch and other saltwater fish—further evidence that the Ubaid people were seafarers.

Outside the temple grounds there was a large cemetery, where the dead lay with their burial articles—pottery vessels, clothing adorned with bands of colored beads, even clay figurines of the gods themselves, personified as male and female entities wearing bitumen headdresses on their lizard-shaped skulls. Each grave was a rectangular shaft, lined with mud bricks. Some form of tomb marker or headstone must have indicated the whereabouts of each grave and the identity of its occupant, for the tombs were often opened—in an orderly manner, with no guesswork—to allow for a second interment. In one case, the secondary burial was not of a spouse or a child, but of a dog. Among the Ubaid dead, there were few signs of any distinction on the grounds of wealth or social rank: no luxuriously furnished tombs or expensive parting gifts to put humbler corpses to shame.

Yet, this apparent egalitarianism failed to survive. Communities grew more populous and more diverse, and the interests of farmers, traders, and artisans could not always coincide. Groups that lived together were now too large to hold—if they ever had—their lands, tools, or other resources in common. Some fields either were more fertile or were closer to the water sources than others; some artisans were better able to gain access to good materials; and some merchants were more successful than their competitors. A new kind of pottery, symptomatic of changing economic relationships, appeared in many Mesopotamian settlements.

These coarse, ugly bowls with beveled rims were manufactured in vast quantities, using a rudimentary means of mass production. Instead of being formed individually on a wheel, they were made of crudely worked clay that had been molded to a uniform shape and size. For some time, archaeologists thought that these vessels might have been the containers for standardized rations of barley, oil, or beer that were doled out to gangs of laborers as wages; however, recent studies have shown that, although the bowls all seem to be the same size, they vary markedly when it comes to the volume of grain or liquid they can hold.

Nevertheless, even though the bowls can no longer be recruited as evidence, this may well have been the point at which society began to divide into employers and employees, into those who owned land and those who worked it. The basic building block of economic and social life may well have been the nuclear family, but this small unit was itself probably part of a clan—an extended family group who combined their efforts to work an ancestral tract of land or who specialized in particular craft skills. Even within this kinship network, certain individuals were likely to have wielded more power, and enjoyed greater authority, than others.

Inequalities of wealth and rank emerged, and they could be seen nowhere more clearly than in the graveyards. The tombs of a rich minority were now resplendent with beautifully crafted metalwork and with beads of ivory and lapis lazuli; the poor, in death as in life, made do with less. But even the lowliest Mesopotamian sought to provide comforts for loved ones in the darkness: A sparsely furnished child's grave near Mosul contained a little flute carved out of bone.

According to archaeological definitions, an ancient site qualifies for the term *city* rather than *town* when it shows evidence of having had the following: a population of thousands rather than hundreds; a central administration, governed by a king, nobles, or priests and housed in temples or palaces; a wide range of commercial and manufacturing enterprises; and a wall around the perimeter to keep out enemies. The first human settlement known to have met all these criteria is the populous city of Uruk, lying on a branch of the Euphrates River 150 miles southeast of the modern-day city of Baghdad. The world had known large settlements before. Çatal Hüyük and Neolithic Jericho were not necessarily unique, but their people left behind only the faintest evidence on the landscape for later ages to discover. When they vanished, so too did their way of life. The inhabitants of Uruk, however, left abundant traces of their lives and labors, and their city gave its name to an important stage in the development of Near Eastern civilization.

Uruk—the biblical Erech—rose to prominence in the second half of the fourth millennium BC, when two adjacent communities, Eanna and Kullaba, expanded and coalesced. Here, between 40,000 and 50,000 people lived in close proximity, in a settlement covering an area of 1,100 acres, which would eventually be enclosed within a six-mile city wall, made of rugged, sun-scorched brick.

Uruk, in common with other Mesopotamian centers, acted as a powerful magnet, and its population swelled at the expense of the smaller settlements in its hinterland. Communities that would once have been self-contained units now became mere tributaries, sending their produce to the city and subordinating themselves to its needs. Settlements that had flourished in the surrounding plain now shrank in population; some were deserted altogether. At the beginning of the fourth millennium BC, at least 146 villages dotted the countryside surrounding Uruk, each with its own temple, irrigation system, and traditional clan-oriented social structure. Within 600 years, that figure had shrunk to just 24, as generations of farmers and their families gradually drifted away to live in the city. In Uruk, these migrants could take up new careers—metalworking, sculpting, brickmaking—which would open up a way of life that was not subject to either the vagaries of the climate or the plundering of hungry

Humans probably used ships to reach Australia around 50,000 BC, but concrete evidence of people traveling by any means other than on foot does not begin until around 3500 BC. At this time, boats such as those shown on page 144 carried goods and animals along the waterways of the Near East. At about the same time, the wheel, perhaps inspired by the use of logs as rollers, transformed humans' ability to transport heavy loads on land. The battle wagon above, depicted on the soundbox of a lyre from a Mesopotamian royal tomb of the third millennium BC, shows that wheels were also valuable in waging war.

This wagon-shape terracotta cup, from a fourth-millennium-BC grave in Hungary, shows that the wheel spread rapidly from its origin in Mesopotamia or Transcaucasia.

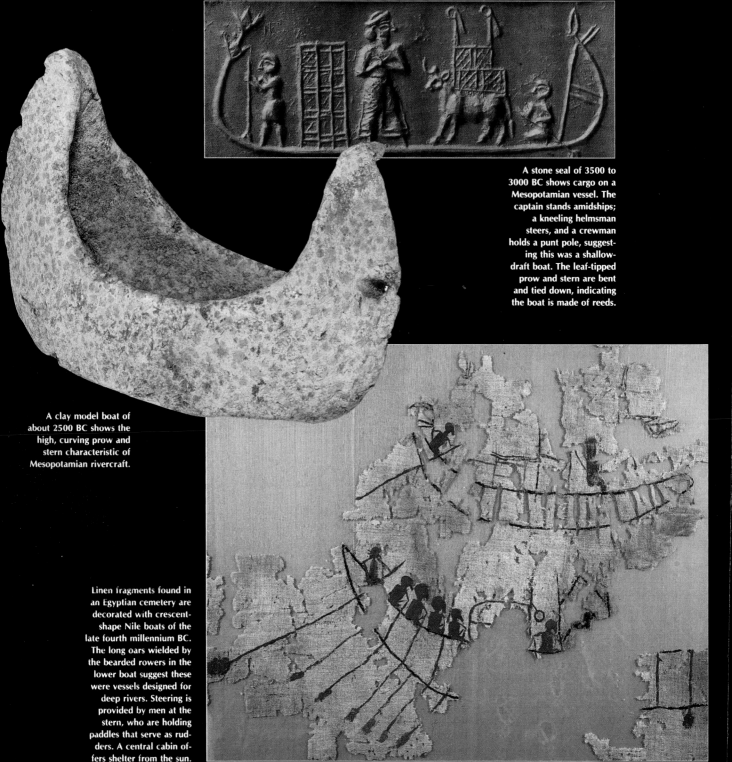

A stone seal of 3500 to 3000 BC shows cargo on a Mesopotamian vessel. The captain stands amidships; a kneeling helmsman steers, and a crewman holds a punt pole, suggesting this was a shallow-draft boat. The leaf-tipped prow and stern are bent and tied down, indicating the boat is made of reeds.

A clay model boat of about 2500 BC shows the high, curving prow and stern characteristic of Mesopotamian rivercraft.

Linen fragments found in an Egyptian cemetery are decorated with crescent-shape Nile boats of the late fourth millennium BC. The long oars wielded by the bearded rowers in the lower boat suggest these were vessels designed for deep rivers. Steering is provided by men at the stern, who are holding paddles that serve as rudders. A central cabin offers shelter from the sun.

nomads who might be driven by jealousy or desperation to attack a prosperous agricultural community.

So began the idea, still prevalent among many communities to this day, that the streets of cities were somehow "paved with gold," and that country dwellers were correspondingly less cultured and less intelligent than their urban counterparts. The first tangible sign of this notion appears in the great Sumerian epic poem *Gilgamesh,* in which a city prostitute taunts the unkempt Enkidu, who has been living in the mountains in preparation for his challenge to Uruk's overbearing King Gilgamesh. Scornfully, the woman asks Enkidu why he chooses to run wild with animals in the hills when he could be in Uruk, where "all the people are dressed in their gorgeous robes, every day is a holiday, and the young men and girls are wonderful to see. How sweet they smell!"

The lands surrounding the city were owned by wealthy men who may have lived either within the town or on their rural holdings, but the actual labor on the farms was done by peasants. These country estates had their own irrigation canals and date-palm orchards; barns to house their draft animals and farming implements; and granaries to store the annual harvest of wheat, barley, sesame, and onions. They were large, well-organized units, possibly consolidating smaller tracts of land, and they may have been held either by clans, or other forms of consortia, or by individual members of these groups.

Within the city itself, there were ample signs of a community that was thriving. Uruk's most well-to-do inhabitants lived in large, two-story houses with wooden balconies that ran the whole length of the upper floor. In these houses dwelled prosperous merchants whose fleets of ships plowed back and forth across the oceans to ports in areas as distant as Bahrain and India. While their sailors were making money for them out on the high seas, the occupants of these grand houses could sit out on their verandas in the evening, drinking beer brought to them by slave women, and look down over the rooftops of less-prosperous neighbors. The latter were a mixture of bureaucrats, merchants, and professional artisans—maybe carpenters, scribes, or weavers—who lived in single-story houses that were less imposing than the adjoining mansions but that nevertheless contained a respectable number of rooms, all opening onto a secluded courtyard, where convivial family gatherings might carry on late into the night.

In the meantime, those with less congenial homes, such as fishermen, water carriers, and unskilled workers, would seek out a comfortable spot to talk over the activities of the day. Even the city's slaves—a mixture of foreign prisoners of war and impoverished rural Sumerians who had sold themselves for the sake of security—may have taken a sip or two from their masters' beer vats as they prepared the evening meal for their households.

Carvings and statues give us some idea of what Uruk's citizens looked like. The men wore flowing, curly beards and went bare chested; from the waist downward hung knee-length, kiltlike garments. The women had their hair braided and oiled, and they covered their bodies in generously cut gowns that concealed everything except their right arms, which were left bare.

The city's prosperity depended on trade. Its merchants may have traveled far from home in search of luxuries such as gold, electrum, ivory, and lapis lazuli, but these were not merely adventurous shopping trips for exotica to please the rich. A regular influx of new goods was crucial for the prosperity of the city as a whole. Uruk could not have survived if the only raw materials available to its skilled workers and traders had been those on hand in the plain around it, and if its customers had been limited to immediate neighbors.

Uruk's distinctive pottery could be found on sites all over Mesopotamia and beyond. So, too, could hundreds of ornately carved, cylindrical stone seals. These bore distinctive and carefully worked designs: plants, animals, strange amalgams of men and beasts, scenes of sacrifice, or symbols of the gods themselves. The seals were pressed into specially prepared surfaces of clay; the clay impressions were then

affixed, by cords, stiff reeds, or strips of cloth, to bundles of merchandise. They served both as marks of ownership and as address labels. Two seal impressions—presumably indicating both vendor and purchaser—were often fastened together on a single bundle to signify a bargain between the parties.

The appearance of these seals along the length of far-flung trade routes indicated the extent of Mesopotamian commercial activity. Most seals were cut from types of stone that could not be found in Mesopotamia, and they turned up as far away as Egypt. It seems that even in the era before the pharaohs ruled, there were active contacts between the peoples along the Euphrates and the Nile rivers.

At the heart of Uruk, rising high above the bazaars and workshops, lay an impressive collection of ceremonial buildings. The first of these, the White Temple, occupied an ancient holy place. It stood on top of the ruins of a much older Ubaid shrine, an arrangement that provides powerful evidence for the existence of a deeply entrenched religious conservatism. New buildings might be erected, and the styles of their decoration might change, but the gods themselves and the rites that honored them were old beyond imagining.

At the White Temple, a handsome triple staircase led up to the sanctuary itself, which was constructed in the same tripartite, cruciform plan that had been used in the region for centuries. Its imposing stepped altar faced a central table used for the placement of burnt offerings; additional staircases gave access to a flat roof, from which one could survey the whole city.

Thirteen hundred feet to the east of the White Temple stood an even larger sacred area, the Eanna compound. This was a complex of broad terraces and courtyards, ceremonial halls, pillared porticoes, and freestanding columns adorned with mosaics of brightly colored terra cotta, glittering in the sunlight like serpents' scales. Underneath these rich facades lay the basic building blocks of Mesopotamia: rectangular mud bricks of uniform size and squared corners, with smooth surfaces that readily lent themselves to elaborate forms of decoration.

The scale of these edifices was monumental. So, too, was the obvious outpouring of wealth, skill, labor, and administrative effort required to build and sustain them. In addition to dominating the landscape, the shrines provided the focal point for every aspect of the city's life. Here, for the first time, was evidence of some form of centralized authority. While some Mesopotamians depended for their daily bread on secular employers and landowners, many others owed their lives and livelihoods to the gods; the temple ruled them.

The gods were the supreme sources of power, the great temples their earthly residences. The priests who carried out ceremonial duties, as well as the scribes, administrators, stewards of farms belonging to the temple, and the artisans and traders at work on the temple's behalf, constituted a god's official household. In form and function, the houses of these divinities may have duplicated the domestic arrangements of ordinary mortals, although on a considerably grander scale. They were divided into private quarters, reception areas, storerooms, and offices, where domestic and public business could be conducted.

The great temples of Mesopotamia, with their manifold ceremonial and economic functions, had evolved from the simple shrines of earlier ages into gigantic ceremonial edifices. At Tell Brak in northeastern Syria, there stood, on top of the town's 130-foot-high mound, an imposing temple dedicated to a cult in which the eye seems

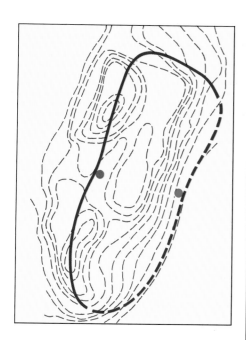

A massive stone watchtower (left) guards the town of Jericho. Almost thirty feet high from top to bottom, the tower was built of rough-hewn stones bonded and faced with mud; access to the flat roof was gained via a steep stairway from an entrance at the base, as shown in the cutaway sections. The plan above shows that the tower, indicated by the red dot, stood just within the town walls, as is thought to have been the case with the abundant, year-round spring (blue dot), although the part of the wall indicated by a thick, broken line has not been excavated. The dotted contour lines indicate the steep incline of the site today—the result of accumulated layers of occupation over several thousand years.

to have been the principal symbol. This impressive structure must have been a source of wonderment to the people of the surrounding Habur plains. Visible for many miles, the Eye Temple's outer walls were covered in skillfully carved rosettes, the petals made of white marble, black shale, and pink limestone. Inside, the whitewashed walls were decorated with more rosettes, interspersed with strips of red limestone inlay and gleaming copper paneling. The inner sanctum housed an altar of colored stone, bordered at the top and bottom by sheet gold that was attached to the stone by gold-headed silver nails. Here, worshipers came by the thousands and left small votive offerings of limestone idols, about two inches high, each with a pair—or pairs—of eyes carved into it.

It was in the south, however, that temple complexes underwent a transformation, becoming virtual cities within cities. The Eanna compound at Uruk, for one example, included not only a temple but also living quarters for a sizable community of priests. In this theocratic society, the priesthood formed a privileged and powerful elite. Its members controlled the religious mysteries, marshaled many of the community's resources, and monopolized skills and knowledge that would be denied to the laity for centuries to come.

On lands owned by the temple, salaried functionaries organized the collection of crops and the distribution of rations among the workers. Those who farmed the gods' fields handed over their produce in the expectation of earning not just earthly rewards but also divine goodwill. Sculpted reliefs show farmers entering the temple, leading sacrificial beasts from their herds and flocks, and carrying baskets brimming with grain, dried fish, and dates.

The building of the shrine itself must have been a massive public-works project. With the experience gained, its priests probably went on to organize other collective efforts, such as the building of communal storehouses or the maintenance of canals. They probably would have employed the town's leading skilled workers, many of whom doubtless welcomed the opportunity to work in the service of the temple; apart from any spiritual favor temple employment might earn them, it would also give them access to the best and rarest materials. The ambitious design and construction of the huge sanctuaries would have both tested the mettle and advanced the techniques of bricklayers and builders, while the task of creating ritual objects of suitable grandeur clearly allowed bronzesmiths and goldsmiths to master all of the metallurgical processes known to the age.

The temple's central position in the economic life of a large and vigorous population also furthered development of systems for keeping records and calculating payments. In order to meet these diverse needs, a new specialist appeared—the *sanga*, a temple official whose title, in the Sumerian language of later generations, meant either priest or accountant. To keep track of the goods that the faithful brought to the temple, the sangas of Uruk attached clay seal impressions to offerings as a means of identifying their donors. They assessed the size of tithes or taxes, and they organized the distribution of rations to the community at large. They needed to know what was owed or promised, how much was given, what the goods were worth in the context of any kind of barter or exchange, and what the temple had to pay out for any services rendered.

The priests who looked after these matters had to become adept at dealing with problems of quantity and value, and they were likely pioneers in the development of Sumerian mathematics, in which the number *60* was the basic unit of calculation. But even the most sophisticated numerical skills could not suffice to keep the institution's affairs in order. The time had come to devise a more efficient means of preserving and communicating complicated information.

For thousands of years, in all parts of the inhabited world, people had found ingenious ways to keep track of simple facts. A shepherd might fill a pouch with pebbles—one stone for each sheep in the flock—or cut notches in a stick to count and remember the beasts in his keeping. The earliest cave dwellers had drawn pictures of humans and animals on the walls of their caverns. These images were the first steps in the gradual evolution of written signs and symbols; they were little more than the barest outlines, giving just enough visual information to communicate a simple factual or ritual message.

In the same way that speech had evolved from the imitation of sounds, writing emerged from the graphic depiction of particular objects. The image that began as a rough but recognizable picture eventually turned into something more abstract: the memory of a memory of a more pictorial sketch. A simple system, based on a collection of descriptive marks, was apparently in use for thousands of years along the trade routes of western Asia. Clay tokens were found as far north as the Caspian Sea, as far south as Khartoum, in Africa, and along an east–west axis extending from the Indus Valley to Anatolia. Over a period of 5,000 years or more, the tokens

On the ivory handle of an Egyptian flint knife *(opposite)*, factions clash in hand-to-hand combat with clubs and maces, while below them, a naval battle rages. The knife dates from around 3400 BC, a time when the rulers of Upper Egypt coveted the more fertile lands of Lower Egypt. A thousand years later, the conduct of war had become a sophisticated discipline, as shown in the images of Sumerian kings leading their troops into battle on the stone slab at right. Mesopotamian city-states were in constant conflict; citizens were often conscripted to swell standing armies, and victories were consolidated by selling prisoners into slavery and installing governors in conquered provinces.

changed very little, suggesting a widespread and enduring network of connections that embraced both the date-palm farmers of southern Iraq and possibly even the original wall builders of Jericho.

By the end of the fourth millennium BC, Mesopotamian priests were using a rudimentary notation system that included many of the same ancient markings, which they incised on small clay tablets for the temple archives. As their communities expanded and transactions grew more complex, so too did the repertoire of signs. In the form of protowriting that emerged in Uruk during this period, pictographs—recognizable images—were gradually joined by the more abstract symbols known as ideograms. By combining the two, the temple scribes were able to represent general ideas, as well as numbers and concrete objects. The outline of a sun, for instance, not only signified the solar orb itself but also the related concept of "day." The sign for an arrow also served as the sign for life. As meanings became more elastic, the signs themselves slowly evolved into a cursive, linear script that was faster and easier to use, but contained little trace of its pictorial origins.

The system also had to be expanded to accommodate a means of designating individual persons. In a small, self-contained community, simple marks, such as those used by potters to identify their work, would have been sufficient. But the number of names used in a city of several thousand people required a more sophisticated solution, and signs representing sounds were gradually incorporated to allow for the phonetic reproduction of different personal names.

The first applications of the art of writing were for financial purposes; only later were these records of accounts and inventories supplemented by lists of gods. By 3000 BC, scribe trainees learned their craft by memorizing a collection of some 1,500 specific signs. These included verbs such as *to buy* and nouns such as *road, expedition, carpenter, smith, copper ingot, donkey, ax, harp, lyre,* and *boat*. There were also signs distinguishing between two- and four-wheeled chariots, and between two different kinds of plow. The vocabulary was not unique to Uruk; the same lists, often written down in the identical sequence, were found in other contemporary Mesopotamian towns. Trade goods and similar pottery styles were no longer the only signs of the interactions between far-flung communities. An articulate, intellectually active society was beginning to emerge.

For centuries, this rudimentary writing would remain the sole preserve of the temples and scribes. Their concerns inevitably dictated the content of the first written materials. But the rich legacy of decipherable texts left by the priests of Sumer made it possible for later generations to acquire a more detailed understanding of their time and place than most other contemporary cultures could supply. And, in an era when social change came slowly, it was likely that this information would have held true for many centuries, both before and after it was written down.

Until the arrival of written records, social hierarchies had been hard to fathom. The men who appeared on carved reliefs, marching in ceremonial processions or performing ritual sacrifices, usually displayed no visible signs of rank. But even in these mute images, certain subtle social differences could be detected. In the carved scenes of lion hunts and harvest festivals, one individual often seemed more important than the rest: He wielded the spear that killed the beast, led the parade of votive offerings, or otherwise stood apart from his companions.

No indication was given as to whether this putative hero was, in fact, a leader of the group or merely a symbolic figure. It was clear, however, that any authority he did possess was still religious rather than secular. He may have been the holder of the first title to emerge in early written texts, sometime around 3000 BC—the *en,* or lord. The en apparently was in charge of the temple establishment. He was as much an administrator as a priest, in charge of the community's economic activities and general welfare. He may well have performed ritual duties within the cult, for the texts in which the en appears suggest that he was responsible for ensuring that the forces of nature remained beneficent: If the harvest was blighted, or the flocks were stricken by disease, the en was obliged to persuade the gods to put things right. Not every en was male. In the city of Ur, at some point early in the third millennium BC, a priestess held the title.

The gods served by the en now began to stand out as individuals, with names, attributes, and histories of their own. The Great Mother was still revered; the Mesopotamians called her Inanna and worshiped her as the goddess of love, fertility, and ripening dates. Her emblem was the gatepost of a temple, and the pictograph of a gatepost became the written symbol of her name. The god Tammuz, depicted in later mythology as Inanna's spouse, represented the generative powers in nature: male animals rutting, or the sprouting seed. Gods and goddesses reigned over the sun and the moon, and they presided over death, warfare, grain, flowing waters, flocks of sheep, and herds of wild asses or gazelles. A goddess particularly beloved by the Sumerians oversaw the fermentation of beer. Small settlements had their own cults, dedicated to obscure local deities.

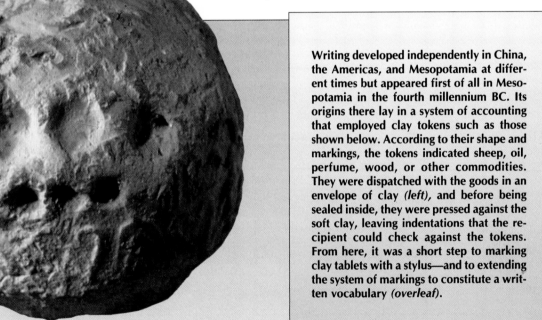

Writing developed independently in China, the Americas, and Mesopotamia at different times but appeared first of all in Mesopotamia in the fourth millennium BC. Its origins there lay in a system of accounting that employed clay tokens such as those shown below. According to their shape and markings, the tokens indicated sheep, oil, perfume, wood, or other commodities. They were dispatched with the goods in an envelope of clay (left), and before being sealed inside, they were pressed against the soft clay, leaving indentations that the recipient could check against the tokens. From here, it was a short step to marking clay tablets with a stylus—and to extending the system of markings to constitute a written vocabulary (overleaf).

FIRST ACCOUNTS

At first, individual gods or goddesses were associated with particular cities. But as these communities expanded and their temples established links among themselves, the diverse histories and often-overlapping identities of the gods became fused—perhaps with conscious intent on the part of a priestly elite—into a single mythological tradition. However, the belief that Sumerian cities were the property of specific gods persisted well into historic times. The temple did not necessarily own or control all the land in the district, but those who did exercise control owed their primary allegiance to the city and its deity—and, by extension, to the deity's representatives. Old tribal allegiances did not disappear but, instead, became subsumed into a larger relationship: that of individuals to their city. Yet there was no word in any of the ancient Mesopotamian languages that survived through written script to express the concept of citizenship. Nor did any of these vocabularies distinguish between village, town, or city.

Mesopotamia had many settlements that grew into cities. Uruk may have been neither the greatest nor the wealthiest among them. According to later tradition, the city of Kish, 100 miles to the north, was a more important economic center, while Nippur, 30 miles south of Kish and in Sumerian mythology home of Enlil, god of air, had greater religious significance.

Inhabitants of these large cities also founded colonies, such as Habuba Kabira in modern-day Syria, about 550 miles north of southernmost Mesopotamia. Habuba Kabira occupied a strategic site, controlling not only the trade routes to the south but also those leading northwest to the Taurus Mountains and southern Anatolia, and due west to the Mediterranean Sea, just 90 miles away. Excavations at Habuba Kabira

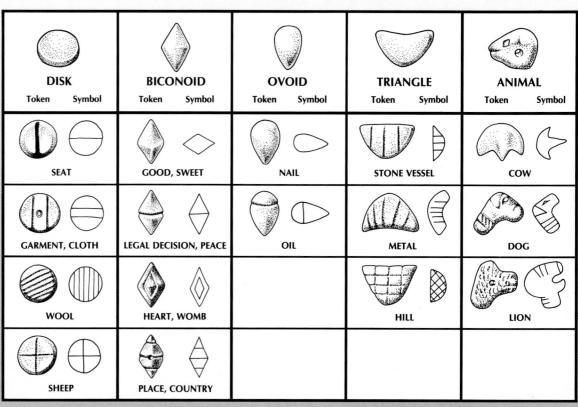

DISK		BICONOID		OVOID		TRIANGLE		ANIMAL	
Token	Symbol	Token	Symbol	Token	Symbol	Token	Symbol	Token	Symbol
SEAT		GOOD, SWEET		NAIL		STONE VESSEL		COW	
GARMENT, CLOTH		LEGAL DECISION, PEACE		OIL		METAL		DOG	
WOOL		HEART, WOMB				HILL		LION	
SHEEP		PLACE, COUNTRY							

show every sign of a settlement that had been founded with a view to dominating the surrounding countryside. A massive fortifying wall, measuring approximately 3,000 feet long and 10 feet thick, with guard towers positioned every 100 feet along its ramparts and a second, smaller wall built in front for good measure, protected three sides of the rectangularly shaped town. On the fourth side stood the broad Euphrates River, Habuba Kabira's main trading artery, along which goods evidently flowed in sufficient quantities to support a large number of businesses and workshops. The civic authorities had enough money to finance the construction of a central administrative and religious complex, along with irrigated gardens, well-ordered streets, and a sophisticated drainage system by which water was drained from houses via clay pipes into a canal network.

The careful town planning and the distinctive, mud-brick building style of southern Mesopotamia show that Habuba Kabira was probably designed and constructed by settlers who had all come from one particular, though as yet unidentified, city in the south. Most other towns and settlements in Mesopotamia did not follow a uniform pattern of development; each one evolved at its own pace, and for different reasons. But, over time, a similar social structure came to prevail in all of them. Sumerian inscriptions written very early in the third millennium BC gave some indication of what this society might have been like.

Temple records listed individual landowners, families due to donate or receive rations, merchants conducting long-distance trade, and lay people carrying out certain administrative responsibilities. The documents indicated that women played an active social role—they owned property, headed households, engaged in foreign commerce, and took their turns holding some form of public office.

The Sumerian clay commodity tokens shown on the far left represented, from top to bottom: cow, stone vessel, sheep, oil, and wool. The chart shows how a token of a particular shape could be marked to convey different meanings, and how the shapes of the tokens were transformed into written symbols that could be incised on a flat tablet.

Two clay tablets from about 3200 BC illustrate successive stages in the development of writing. The one above is probably a list of goods, with both impressed markings and inscribed characters. Only the latter appear on the tablet at right, and although their meaning has not been fully deciphered, they express a more complex message.

Slowly, the balance of power began to shift away from the temple and toward secular government; it is likely that temples gradually began to share power with groups of clan elders who came together in some form of local assembly. Early Sumerian myths described councils of the gods, which bear a strong resemblance to existing political institutions. The members of these divine bodies passed judgments, carried out sentences, and held emergency meetings in times of crisis.

Gradually, individual leaders emerged. The lord of the temple, the en, was now joined by two other titleholders: *lugal* and *ensi*. The term *lugal* meant some kind of great man or master; the contexts in which it was used made it unclear whether lugal was a general term for an influential, prestigious public figure or an actual king. Lugals apparently acted as leaders, but they may have combined both secular and religious functions, since they were involved in the administration of the temple and the performance of certain rites within the cult.

The title *ensi* sometimes appeared in the same context as lugal, and it was occasionally used in a way that suggested that ensi may have been an alternative title—applied in some other city—for a person of the same high rank. Other texts used ensi to mean a figure who, while still lordly, was subordinate to some higher master. By the middle of the third millennium BC, the ensi was the title of a city governor appointed by some higher authority—which might have been the temple priests, the assembly of elders, or a lugal who had become a power in the land.

Secular leaders now started to assert their place in the Mesopotamian social structure. The earliest Sumerian myths and legends sketch a society where individuals began to claim royal status, either because they had been heroes who founded a city or because they had been nominated by the gods. Subsequent Sumerian lists of kings hark back to leaders of some unspecified prehistoric period, such as Enmerkar, calling him "the man who built Uruk," or refer to a tradition that justifies kingly status by divine appointment. A monarch named Etana, for instance, is cited as the founder of the dynasty that ruled the great third-millennium city of Kish; he was assigned by the gods to seek out the "plant of birth," the source of hereditary royal power, as a means of introducing the institution of kingship to the human race. As a way of countering possible challenges to their power, the kings were careful to stress their divine patrons or parentage, conferring upon themselves such titles as "the beloved son of the goddess Ninhursag."

Relationships between these emergent kings and the old convocations of city elders may have been precarious. Gilgamesh, the Mesopotamian hero of the world's earliest-known epic, claimed that the gods had appointed him king of Uruk. Nevertheless, when he wanted to go to war against his enemies, he was compelled to seek the approval of the city's elders. They refused to grant it, so he took matters into his own hands and replaced them with a convocation of younger men, who gladly authorized him to wage war.

The need for military leadership brought individual rulers such as Gilgamesh into the limelight. Even if a council of elders reigned supreme, they would have had to appoint some individual to lead their soldiers into battle. The rise of the individual, secular king may thus have been a reflection of the increased—and sustained—hostility between communities. Cities looked beyond their borders; they were no longer content to rule only over their hinterlands. And, as a region prospered, its settlements promised rich spoils for an invader. The war leader previously appointed by an assembly to deal with a temporary emergency now held his command on a more permanent basis. If he succeeded in keeping a city's enemies at bay, or in increasing the territory under its control, he might found a dynasty, ensuring that his sons would rule after him.

The carved reliefs, stone seals, and burial monuments of the era spoke of a new obsession with warfare. Pictures of hunting trips or peaceful harvest festivals were joined by bloody battle scenes. The funerary items of the age reflected this new concern with military matters: Men of evident importance were now buried not just with an arsenal of shining weapons but also with the bodies of their armed retainers, put to death on their master's demise. The metalworkers would find the soldier as loyal and lucrative a customer as the priest. In the same period, the written sign for a slave—derived from the word for a foreign country—entered the scribal lexicon, while a seal from Uruk showed images of bound captives being marched off after a battle. In such a climate, it was no wonder that the inhabitants of Gilgamesh's city organized labor gangs to erect and maintain a six-mile-long wall.

The first steps toward an urban society in the Near East—defended settlements, far-reaching trade networks, irrigation systems, dominant religious institutions, specialized crafts, and incipient social and political hierarchies—came earlier than their equivalents in other parts of the world. But the builders of Jericho, the devout artists of Çatal Hüyük, the prosperous Ubaid farmers, and Uruk's priestly elite were not unique. Far to the east, in the valley of the great Indus River that flowed down from the Himalayas, and beyond those mountains in China's vast expanses, their contemporaries traveled their own distinctive paths toward civilization.

It was in the Kachhi plain, between the bare hills of Baluchistan in modern-day Pakistan, and the fertile Indus Valley in northwest India, that the earliest flowering of Indian civilization occurred. The first people to inhabit the Kachhi plain were nomads from the Baluchistan mountains, who would descend each winter in order to escape the merciless cold that swept the crags. Throughout the months that snow was coating their homelands, they were able to graze their herds on the lush Kachhi pastures and grow cereal for themselves on land they had fertilized with silt from dammed-up streams. Gradually, more and more nomads began to stay on when summer came, rather than retreat to the hills; better to be hot in a land of plenty, they reasoned, than to be cool in a land of shortage.

Between 8000 and 5000 BC, people on the western fringes of the Indus Valley system built up prosperous and long-lasting farming settlements. By the end of the sixth millennium BC, they were living in mud-brick houses, cultivating wheat and barley, and tending their thoroughly domesticated sheep, goats, and cattle. So organized had their societies become that the village of Mehrgarh, built on a terrace overlooking the Bolan River, had a collection of storehouses that probably acted as central distribution points for food, craft materials, and manufactured goods.

Like their Near Eastern contemporaries, the peoples of the Indus Valley fringe learned how to manage the floods of the river so as to keep their crops watered. The resulting prosperity allowed them to exploit to the fullest their geographical position at the center of a number of converging trade routes.

Their communities were small, but they had an appetite for exotic luxuries. They wore turquoise beads that may have come from as far north as Turkmenia, in central Asia, and conch shells from the Arabian Sea, some 300 miles away. Among their prized possessions were pendants of perforated lead and a few beads made from

Like tombstones in a vast burial ground, the standing stones of Carnac stretch across the Breton countryside. This is a part of the site in which rough granite pillars, some of them more than twelve feet high, fan from east to west in twelve straight rows.

Over a period of 2,500 years, beginning in the fifth millennium BC, the first farmers of Europe marked their presence with ritual stone monuments, constructed on a stupendous scale. At Carnac, in northwestern France, approximately 3,000 stones remain standing out of an original 10,000. Some of the stones are arranged in egg-shape rings, others in straight lines that stretch for several hundred yards. Their alignment is thought to chart in some way the courses of the sun and moon.

THE MEGALITHS OF CARNAC

A stylized female outline *(right)* guards the entrance passage of a burial chamber at Carnac. Below, spiral carvings, perhaps reflecting lunar or solar orbits, adorn a passageway at Gavrinis, off Brittany.

copper, which must have been a rarity, brought from some remote and unknown source. They would not develop their own metallurgical skills or make any significant use of copper tools until the beginning of the fourth millennium BC.

They placed the most desirable of their trade goods in the graves of the dead, along with human figurines of unbaked clay; food offerings were stored in reed baskets, coated with a protective layer of bitumen that made them as watertight as clay. As yet, the craft of pottery was still unknown to them, but when the artisans of the region did begin to work in ceramics, early in the fifth millennium BC, their designs and decorations would echo Mesopotamian styles so strongly that the resemblances may well have been more than coincidental.

The peoples of the upland valleys and the flood plains below them seem to have engaged in a constant flow of commercial and other exchanges. Communities may not have been large, but neither were they isolated or provincial. A large number of settlements grew and prospered because of their location at logical stopping places on longer trade routes. Caravans of merchandise, stopping to rest or to engage in barter, might swell the population of a settlement for only brief periods, leaving no traces more permanent than the marks of their tent posts. But they became frequent and regular visitors, as reliable a source of wealth as any farmer's crops. To make their journeys possible, the traders needed these towns as much as the local inhabitants needed the merchants. Other settlements might come and go, but the sites along the caravan routes prospered; merchants transporting trade goods through the western Indus Valley in 2000 BC passed through towns that had thrived continuously for more than 1,500 years.

To the west of the Indus Valley, towns such as Mundigak, in present-day Afghanistan, demonstrated movements toward an urban revolution that bore many resemblances to developments in the Near East. The town, on a well-trodden trade route into the heart of central Asia, would have witnessed the passage of lapis lazuli, from the Afghan hills down to the coast, on its long journey to the jewelers of Sumer. Mundigak was already ancient before the third millennium BC began; its inhabitants lived in well-built houses with hearths located at the center of many rooms. They had originally used stone to make tools and weapons, carving leaf-shape arrowheads, but by 3000 BC, they were taking an interest in metallurgy, working in copper and—by mid-millennium—in bronze. They loved vivid decoration. Their wheel-thrown pottery vessels glowed with painted polychrome designs—naturalistic scenes of animals, foliage, and swimming fish, or handsome black geometric motifs on a deep red background. They carved flat stone seals into intricate shapes and patterns—zigzags, mazes, diagonal grids, and interlacing curves—using them, presumably, to print decorative designs on their woven cloth.

By about 2500 BC, Mundigak had turned into a substantial city. And, like the great urban centers of Mesopotamia, it boasted a large temple complex rising above the town. In Mundigak, however, the temple complex had a secular rival—an imposing brick palace, flanked by a colonnade of rounded pillars. But Mundigak's long and increasing prosperity had excited envy. In response to real or imagined threats of invasion, the entire town was now protected by imposing walls and square watchtowers, solidly constructed of sun-dried brick.

As the region's population swelled and new settlements sprang up in all directions along the Indus River system, the interactions between these communities grew ever stronger. Trade may not have been the only reason for the spread of skills, such as pottery or metalworking, and the disappearance of local differences in decoration in favor of more uniform styles.

As the material achievements of the Indus Valley culture extended over an enormous geographical area, so, too, did signs of its spiritual beliefs. For hundreds of years, female figurines had been fashioned in changing materials and styles, from rough shapes of unbaked clay to elaborately coiffed terra-cotta figurines. Soon after 3000 BC, they were joined by animal companions: bulls and buffaloes, with six-petaled flowers or pipal-leaf garlands growing up between their horns. These deities, or their descendants with strikingly similar attributes, would be worshiped throughout the Indian subcontinent well into modern times. Their devotees might be overrun or absorbed by waves of invaders or more peaceable migrants, but they survived as proof of the extraordinary breadth and durability of spiritual seeds first planted in the flood plain of the Indus River system.

Those seeds were to bear their richest fruit in the form of the first great Indian civilization that would flourish in the Indus Valley throughout the second half of the third millennium BC. With its two great cities of Harappa and Mohenjo-Daro, its strict, rectilinear street plans, and its use of uniform weights and bricks, this society owed more than a little to the tradition of well-ordered Indus Valley communities that stretched back more than 5,000 years.

Between 7000 and 5000 BC, small farming communities scrabbled for a living in isolated clusters along the banks of China's great rivers. There were few, if any, signs of contact among them. The northerners lived in semisubterranean huts made of wattle and daub, and they cultivated millet, which they stored in pits. In the south, a landscape pockmarked with caves provided many natural dwellings; however, the current occupants had little in common with their Paleolithic ancestors; they still hunted and fished for much of their food, but they had also learned how to grow rice.

In territories vulnerable to floods and other natural disasters, the lives of southerners and northerners alike must have been precarious. They became early experts in scapulimancy, a fortunetelling technique based on the application of heat to tortoise shells or animal bones and the subsequent study of the resultant cracks. The pattern of fractures was read like a map of the future, to see what lay ahead. To interpret these mysteries more precisely, diviners carved various symbols on the bones or shells before subjecting them to the fire. Inscriptions of this kind, made in a settlement in present-day Henan province, sometime around 6000 BC, may have been the earliest pictographs, precursors of the Chinese characters that would not come into use until the second millennium BC. If so, the fortunetellers of Henan predated the achievements of Mesopotamia's scribes by at least 2,000 years.

In the centuries between 4000 and 3000 BC, these disparate communities began to coalesce. Their inhabitants were no longer simply residents of a single village or members of a clan but part of a larger sphere of cultural contacts. During the next millennium, the Chinese, too, began to live in large, well-defended towns, in a society of specialized artisans and strongly differentiated social classes—a process of centralization that would eventually culminate in the establishment of China's first ruling dynasty, the Xia, soon after 2000 BC.

The inhabitants of Chengziyai, a populous community in Shandong province, lived in troubled times. They possessed an arsenal of bows and arrows that were clearly designed for warfare rather than simple hunting, and they sought to protect themselves against some external threat by building a rampart of earth twenty feet high and thirty feet wide. Such a substantial structure suggested that the town had a large labor force at its disposal, yet it was unlikely that all inhabitants were equally obliged to dirty their hands: The graves in the town's cemetery indicated marked differences of wealth and status.

Most of the burials were simple, spartan affairs, in which the corpse was placed, uncoffined, in a narrow pit. The second largest group of burial sites included a few simple burial articles. Nearby, however, were the tombs of their social superiors, whose bodies were buried in caskets, accompanied by jade ornaments and beautifully crafted pottery. These, however, were outshone by the graves of an even more exclusive group: Their tombs were capacious, equipped with ledges and other architectural features and lavishly endowed with luxury goods, including a curiously wrought, long-stemmed goblet, mounted on the carved jawbone of a pig.

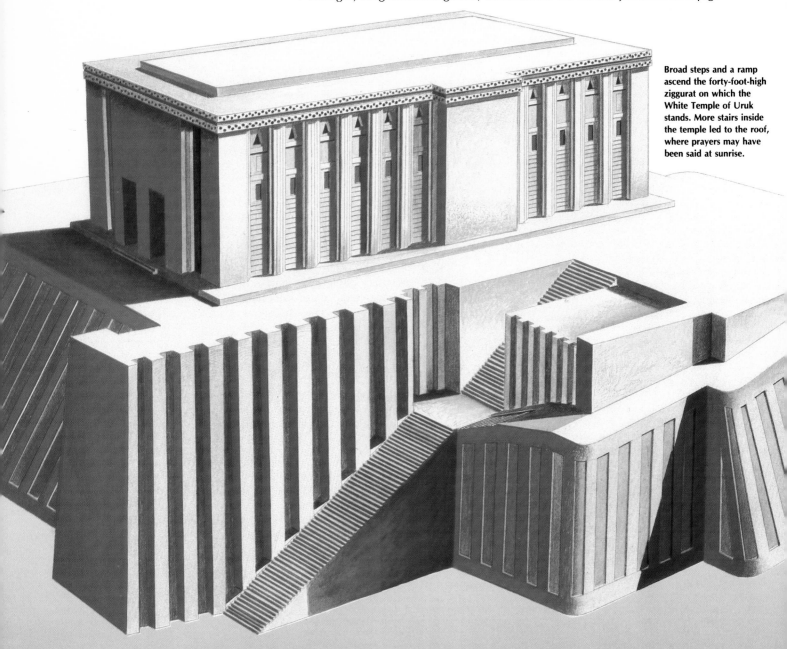

Broad steps and a ramp ascend the forty-foot-high ziggurat on which the White Temple of Uruk stands. More stairs inside the temple led to the roof, where prayers may have been said at sunrise.

Funerary items of the period suggested an elaborate system of religious beliefs. Ritual objects found in burial sites included oracle bones, like those used by the fortunetellers of seventh-millennium Henan, and realistic bird and animal masks. Some well-furnished graves also contained secondary burials of additional skulls and limbs; these may have been relatives of the original corpse, buried later, or sacrificial victims dispatched to serve as attendants in some form of afterlife. Human sacrifice may also have been practiced in an attempt to benefit the living, particularly the members of the emergent upper classes; certain dwellings, better built and more substantial than the neighboring houses, contained infants buried in urns under the foundations or inside the walls.

The artisans who created the goods for this prototypical aristocracy were sophisticated practitioners of their various arts. Potters knew how to produce eggshell-thin drinking cups, matching jars, and lidded boxes, all finished with a lustrous black glaze. Metalworkers had mastered the mysteries of copper—and, by the end of the third millennium BC, bronze. Stonecutters had achieved equal heights of skill in the treatment of jade. Different varieties of jade were imported from widely scattered sources to achieve a range of visual effects; the stone was cut into fine rings or carved into tubes—square on the outside, round inside—incised with elaborate surface decorations.

The graves of the era gave evidence, too, of a less glorious development. In the village of Jiangou in present-day Hebei province, at some point in the course of the third millennium BC, dozens of corpses were dumped unceremoniously into a well. Their skulls were dented or shattered by heavy blows; some bore the marks of posthumous scalping; other heads were missing altogether, apparently removed and carried off as trophies. The position of some remains suggested that a few people had been buried alive and had struggled vainly to climb out of the pit. Stone arrowheads and spearheads lay scattered around the site, near the ruins of an earthen wall. Brutal murder was nothing new; people

In a procession around this alabaster vase from Uruk, naked priests offer the first fruits of the land to Inanna, goddess of love and fertility. At the top of the vase, she receives a jar brimming with produce.

THE SEAT OF PRIESTLY POWER

Dominating the skyline of the early cities of Mesopotamia, massive temples proclaimed the importance of the local deities—and of the exclusive class of priests and priestesses who mediated between the people and their gods. The temple shown in the drawing at the far left was constructed around 3200 BC at Uruk, a city that stood on the Euphrates River about 150 miles southeast of the modern-day city of Baghdad. The high platform, or ziggurat, contained the bricked-in remains of earlier temples that had stood on the site, which was dedicated to the god of the sky, Anu.

Frequently decorated with elaborate clay mosaics colored in black, red, and white, these formidable buildings impressed upon the common people the political as well as the spiritual power of organized religion.

had slaughtered each other since time began. But the carnage at Jiangou provided the first concrete evidence of mass violence in Chinese history.

In Hebei as in Jericho, on the banks of the Indus as on the banks of the Euphrates, men had become adept in yet another skill: the systematic waging of war. Some 7,000 years of progress on the part of the farmer, the smith, the trader, and the temple priest had led, however indirectly, to this development. The farmer had learned how to exploit and enhance the wealth of the land, making it valuable enough for others to covet. The bronze worker had provided an arsenal of portable, efficient, metal weapons, capable of far more devastating damage than cumbersome clubs or missiles made of stone. The trader had beaten paths across unknown territory, establishing routes for invading armies to follow. And the priests of the temples had shown how to marshal large forces for collective enterprises, how to organize their labors, and how to distribute the rations and keep them at work.

Men would wage war for straightforward reasons: to claim new territory, to win control over water sources, to expand the power of their own group. The emerging social elites that dominated the great towns and their hinterlands now looked to extend their domains. Cities turned into city-states, which would themselves, in time, turn into nations and empires. Yet not even the greatest of these was immune to threats from more powerful invaders. The proudest conquerors were inevitably conquered in their turn.

Civilization endured a bloody birth. The earliest written narratives recounted battles; the first individual names and faces to emerge into the light of history were those of military heroes and warrior-kings. But ordinary people had greater and more peaceable victories to their credit. In the course of a few thousand years, they had gained some form of power over the world they lived in. Whereas they had once been Nature's passive victims, they became its collaborators or, indeed, its conquerors. They had learned how to farm, how to exploit the earth's mineral resources, how to make tools that would extend their physical capabilities, how to divide their labors to make fullest use of these new skills, and—most important of all—how to communicate and preserve their expanding knowledge. In so doing, they drew the blueprint for the modern world.

5,000,000-3,000,000	3,000,000-2,000,000	2,000,000-1,000,000	1,000,000-500,000

EUROPE

First hominids reach Europe from Africa (850,000 BC).

THE AMERICAS

AFRICA

Australopithecines, the earliest-known bipedal hominids, inhabit East Africa's Great Rift Valley and southern Africa.

First evidence of *Australopithecus afarensis* in eastern Ethiopia, dating from 3.5 million BC.

Australopithecus africanus inhabits southern Africa. First evidence in eastern and southern Africa of *Homo habilis,* who made use of stone tools (2.3 million BC).

Australopithecus robustus inhabits southern Africa. Evidence of *Australopithecus boisei* at Olduvai Gorge in Tanzania.

Homo erectus emerges in Africa around 1.6 million BC.

THE NEAR EAST

Western Asia is inhabited by hominids.

THE FAR EAST

TimeFrame 5,000,000-3000 BC

r-gatherers live in seasonal
ments such as that at Bil-
eben in Germany.

nce of human use of fire at
Amata, southwestern
e.

Homo sapiens neanderthalensis emerges in Europe.

Homo sapiens sapiens now present in Europe.

Earliest evidence of human settlement at Pedra Furada in Brazil.

Earliest evidence for *Homo sapiens sapiens*, anatomically modern humans, in southern Africa.

Homo sapiens neanderthalensis present in regions from the Levant to central Asia.

s at Zhoukoudian in China
occupied by humans. Evi-
e of the use of fire for
king.

Humans reach Australia by boat from Indonesia.

30,000-20,000	20,000-15,000	15,000-10,000	10,000-9000	9000-8000

EUROPE

Cave paintings and Venus figurines bear witness to new artistic skills.

In southeast Europe and West Asia, tools and ornaments are made from beaten copper.

THE AMERICAS

Humans reach the southern tip of South America. Both North and South America inhabited by big-game hunters.

AFRICA

Grindstones are used in Upper Egypt and Nubia to make flour from wild grass seeds.

Beginning of Saharan rock ar including paintings of animal on rocks and in shelters.

THE NEAR EAST

Sheep domesticated in northe Mesopotamia.

THE FAR EAST

First evidence of pottery at the Fukui cave in Japan.

00-7000	7000-6000	6000-5000	5000-4000	4000-3000
	Agriculture develops in the Balkans and the west and central Mediterranean regions. Sheep, goats, and cereals are domesticated.	Agriculture spreads as far north as the Netherlands. Technique of smelting is developed to extract copper from ore.	First megalithic monuments in northwestern Europe.	The simple plow is used in northern and western Europe. Evidence of wheeled vehicles in eastern Europe. Agricultural settlement founded at Skara Brae in the Orkneys.
ps are cultivated in Mexico.	Crops are cultivated in Peru.			Earliest evidence of domestication of corn in Tehuacán Valley, Mexico. Domestication of the guanaco, a llama-like animal.
		Domestication of native North African cattle in the Saharan region.		
y domesticated wheat, bar- and beans cultivated in the le Crescent.	Wheat, barley, and beans cultivated from Anatolia to Pakistan. Domesticated animals include goats, pigs, and cattle. First evidence of textiles. Obsidian is widely traded in the Middle East. Çatal Hüyük, one of the first settled towns, is founded in Anatolia.	Evidence of the earliest-known irrigation system in Mesopotamia. Northern Mesopotamia and Iran become centers of trade for painted pottery.	First evidence of the human use of sail in Mesopotamia. Domestication of the horse on The Steppes.	The world's first urban civilization emerges in southern Mesopotamia. First evidence of the use of boats and wheels in the Near East. Kilns and turntables are used by Mesopotamian potters. In Mesopotamia, evidence of the earliest writing.
		Farming villages established in China. Millet is cultivated, and pigs and dogs are raised. Wet-rice farming established at Hemudu near the east coast of China.		Millet is cultivated in Korea.

ACKNOWLEDGMENTS

The editors wish to thank the following individuals and institutions for their valuable assistance in the preparation of this volume:

England: Bristol—Bob Savage, Professor of Vertebrate Palaeontology, University of Bristol. Cambridge—R. A. Foley, Lecturer in Biological Anthropology and Fellow of King's College; Jane M. Renfrew, Department of Archaeology, University of Cambridge; T. E. G. Reynolds, Evans Fellow, University of Cambridge; Sander van der Leeuw, Department of Archaeology, University of Cambridge. Hull—Paul Bahn. London—Roderick Conway Morris; Harriet Crawford, Institute of Archaeology, University College; Andy Current, The Natural History Museum; Liz Hodgson; Noah Klein; Ferdie McDonald; Caroline Manyon; James Mellaart, Institute of Archaeology, University College; Christine Noble. Southampton—Clive Gamble, Department of Archaeology, University of Southampton.

France: Paris—François Avril, Curateur, Département des Manuscrits, Bibliothèque Nationale; Hélène du Saussois, Département des Antiquités Orientales, Musée du Louvre.

Hungary: Budapest—Pal Raczky, Department of Archaeology, Budapest University; Laszlo Selmetszy, Director, Törtely Museum, Budapest.

U.S.A.: Texas—Denise Schmandt-Besserat, Professor of Middle Eastern Studies, University of Texas at Austin. Wyoming—George C. Frison, Professor of Anthropology, University of Wyoming.

PICTURE CREDITS

BIBLIOGRAPHY

BOOKS

Aitchison, Leslie, *A History of Metals.* Vol.1. London: Macdonald & Evans Ltd., 1960.

Bacon, Edward, ed., *Vanished Civilizations: Forgotten Peoples of the Ancient World.* London: Thames and Hudson, 1963.

Bahn, Paul G., and Jean Vertut, *Images of the Ice Age.* London: Windward, 1988.

Baltscheffsky, H., H. Jornval, and R. Ringler, eds., *Molecular Evolution of Life.* Cambridge: Cambridge University Press / Royal Swedish Academy of Sciences, 1986.

Barker, Graeme, *Prehistoric Farming in Europe.* Cambridge: Cambridge University Press, 1985.

Barrow, John D., and Frank J. Tipler, *The Anthropic Cosmological Principle.* New York: Oxford University Press, 1988.

Bass, George F., ed., *A History of Seafaring Based on Underwater Archaeology.* London: Thames and Hudson, 1972.

Bender, Barbara, *Farming in Prehistory: From Hunter-Gatherer to Food-Producer.* London: John Baker, 1975.

Benton, Michael, *Dinosaurs: An A-Z Guide.* London: Kingfisher Books, 1988.

Bordaz, Jacques, *Tools of the Old and New Stone Age.* Newton Abbot: David & Charles, 1971.

Bordes, François, *The Old Stone Age.* Transl. by J. E. Anderson. London: Weidenfeld and Nicolson World University Library, 1968.

Breuil, Abbé H., *Four Hundred Centuries of Cave Art.* Transl. by Mary E. Boyle. Montignac, Dordogne: Fernand Windels, 1952.

Brodrick, A. Houghton, ed., *Animals in Archaeology.* London: Barrie & Jenkins, 1972.

Burl, Aubrey:
Megalithic Brittany: A Guide to over 350 Ancient Sites and Monuments. London: Thames and Hudson, 1985.
Rings of Stone: The Prehistoric Stone Circles of Britain and Ireland. London: Weidenfeld and Nicolson, 1979.

Chang Kwang-chih, *The Archaeology of Ancient China.* New Haven: Yale University Press, 1986.

Charleston, Robert J., ed., *World Ceramics.* London: Paul Hamlyn, 1968.

Chia Lan-po, *The Cave Home of Peking Man.* Beijing: Foreign Languages Press, 1975.

Childe, V. Gordon, *Skara Brae: A Pictish Village in Orkney.* London: Kegan Paul, Trench, Trubner, 1931.

Clarke, David, and Patrick Maguire, *Skara Brae: Northern Europe's Best Preserved Prehistoric Village.* Scotland: Historic Scotland, 1989.

Clutton-Brock, Juliet, *Domesticated Animals from Early Times.* London: Heinemann and British Museum (Natural History), 1981.

Cox, Barry, Dougal Dixon, Brian Gardiner, and R. J. G. Savage, *Illustrated Encyclopedia of Dinosaurs and Prehistoric Animals.* London: Macmillan, 1988.

Craig, Annabel, *The Usborne Book of Prehistoric Facts.* London: Usborne Publishing, 1986.

Davis, Simon J. M., *The Archaeology of Animals.* London: B. T. Batsford, 1987.

Dennell, Robin, *European Economic Prehistory: A New Approach.* London: Academic Press, 1983.

Deutsches Archäologisches Institut, *Denkmäler Antiker Architektur.* Berlin: Verlag Walter De Gruyter, 1982.

Dixon, J. E., J. R. Cann, and Colin Renfrew, "Obsidian and the Origins of Trade," from *Avenues to Antiquity.* San Francisco: W. H. Freeman, 1976.

Edwards, I. E. S., C. J. Gadd, and N. G. L. Hammond, eds., *Prolegomena and Prehistory.* Vol. 1, Part 1 of *The Cambridge Ancient History.* Cambridge: Cambridge University Press, 1970.

Fagan, Brian M.:
The Great Journey: The Peopling of Ancient America. London: Thames and Hudson, 1987.
New Treasures of the Past: Fresh Finds That Deepen Our Understanding of the Archaeology of Man. London: Windward, 1987.

Foley, Robert, *Another Unique Species: Patterns in Human Evolutionary Ecology.* Harlow: Longman Scientific & Technical, 1987.

Foley, R., ed., *Hominid Evolution and Community Ecology: Prehistoric Human Adaptation in Biological Perspective.* London: Academic Press, 1984.

Frison, George C., ed., *The Casper Site: A Hell Gap Bison Kill on the High Plains.* New York: Academic Press, Inc., 1974.

Gamble, Clive, *The Palaeolithic Settlement of Europe.* Cambridge: Cambridge University Press (Cambridge World Archaeology), 1986.

Gimbutas, Marija, *The Gods and Goddesses of Old Europe: 7000 to 3500 B.C.: Myths, Legends and Cult Images.* London: Thames and Hudson, 1974.

Goudie, A. S., *Environmental Change.* Oxford: Clarendon Press, 1983.

Gowlett, John, *Ascent to Civilization.* London: Collins, 1984.

Graziosi, Paolo, *Palaeolithic Art.* London: Faber and Faber, 1960.

Greenwood, P. H., *A Living Fossil Fish: The Coelacanth.* London: British Museum (Natural History), 1988.

Hall, H. R., C. Leonard Woolley, et al., *Al-'Ubaid.* Vol. 1 of *Ur Excavations.* Oxford: Oxford University Press, 1927.

Halstead, L. B., *Hunting the Past: Fossils, Rocks, Tracks and Trails: The Search for the Origin of Life.* London: Hamish Hamilton, 1982.

Hamblin, Dora Jane, and the Editors of Time-Life Books, *The First Cities.* (The Emergence of Man series). New York: Time-Life Books, 1973.

Harding, Dennis, *Prehistoric Europe.* (The Making of the Past series). Oxford: Elsevier-Phaidon, 1978.

Hawkes, Jacquetta, *The Atlas of Early Man.* London: Macmillan, 1976.

Hay, Richard L., *Geology of the Olduvai Gorge.* Berkeley: University of California Press, 1976.

Heinrich, Ernst, *Die Tempel und Heiligtümer Im Alten Mesopotamien: Typologie, Morphologie und Geschichte.* Berlin: Verlag Walter De Gruyter, 1982.

Howell, F. Clark, and the Editors of Time-Life Books, *Early Man.* (Life Nature Library series). Hong Kong: Time-Life Books, 1984.

Howells, William W., "Homo Erectus," from *Avenues to Antiquity.* San Francisco: W. H. Freeman, 1976.

Huyghe, René, ed., *Larousse Encyclopedia of Prehistoric and Ancient Art.* London: Paul Hamlyn, 1962.

Imbrie, John, and Katherine Palmer Imbrie, *Ice Ages: Solving the Mystery.* London: Macmillan, 1979.

Institute of Vertebrate Paleontology and Paleoanthropology, Chinese Academy of Sciences, *Atlas of Primitive Man in China.* Beijing: Science Press, 1980.

Jelínek, J., *The Pictorial Encyclopedia of the Evolution of Man.* London: Hamlyn, 1975.

Johanson, Donald C., and Maitland A. Edey, *Lucy: The Beginnings of Humankind.* London: Granada, 1981.

Kenyon, Kathleen M., *The Architecture and Stratigraphy of the Tell.* Vol. 3 of *Excavations at Jericho.* London: British School of Archaeology in Jerusalem, 1981.

Klein, Richard G., "Ice-Age Hunters of the Ukraine," from *Avenues to Antiquity.* San Francisco: W. H. Freeman, 1976.

Kulke, Hermann, and Dietmar Rothermund, *A History of India.* London: Croom Helm, 1986.

Kurten, Björn, *Pleistocene Mammals of Europe.* London: Weidenfeld and Nicolson, 1968.

Lamb, H. H., *Climate, History, and the Modern World.* London: Methuen, 1982.

Leakey, Mary, *Disclosing the Past.* London: Weidenfeld and Nicolson, 1984.

Leakey, Richard E.:
The Making of Mankind. London: Michael Joseph, 1981.
One Life: An Autobiography. London: Michael Joseph, 1983.

Leakey, Richard, and Roger Lewin, *People of the Lake—Man: His Origins, Nature and Future.* London: Penguin Books, 1978.

Leroi-Gourhan, André:
"The Evolution of Paleolithic Art," from *Avenues to Antiquity.* San Francisco: W. H. Freeman, 1976.
Préhistoire de l'Art Occidental. Paris: Éditions d'Art Lucien Mazenod, 1965.

Lumley, Henry de, "A Paleolithic Camp at Nice," from *Avenues to Antiquity.* San Francisco: W. H. Freeman, 1976.

Manley, John, *Atlas of Prehistoric Britain.* Oxford: Phaidon, 1989.

Marshack, Alexander, *The Roots of Civilization: The Cognitive Beginnings of Man's First Art, Symbol and Notation.* London: Weidenfeld and Nicolson, 1972.

Mellaart, James:
Çatal Hüyük: A Neolithic Town in Anatolia. London: Thames and Hudson, 1967.
"A Neolithic City in Turkey," from *Avenues to Antiquity.* San Francisco: W. H. Freeman, 1976.

Moody, Richard, *Prehistoric World.* London: Hamlyn, 1980.

Oates, David, and Joan Oates, *The Rise of Civilization.* (The Making of the Past series). Oxford: Elsevier-Phaidon, 1976.

Past Worlds: *The Times Atlas of Archaeology.* London: Times Books, 1988.

Piggott, Stuart:
Ancient Europe from the Beginnings of Agriculture to Classical Antiquity. Edinburgh: Edinburgh University Press, 1965.
"The Beginnings of Wheeled Transport," from *Avenues to Antiquity.* San Francisco: W. H. Freeman, 1976.
The Earliest Wheeled Transport: From the Atlantic Coast to the Caspian Sea. London: Thames and Hudson, 1983.

Piggott, Stuart, ed., *The Dawn of Civilization: The First World Survey of Human Cultures in Early Times.* London: Thames and Hudson, 1961.

Pinna, Giovanni:
The Dawn of Life. (The World of Nature series). London: Orbis Publishing, 1972.
Prehistory. London: Burke Publishing, 1985.

Prehistoric Times: *Readings from Scientific American.* San Francisco: W. H. Freeman, 1983.

Reader, John:
Missing Links: The Hunt for Earliest Man. London: Collins, 1981.
The Rise of Life: The First 3.5 Billion Years. London: Collins, 1986.

Renfrew, Jane M., "The Archaeological Evidence for the Domestication and Exploitation of Plants: Methods and Problems," from *The Domestication and Exploitation of Plants and Animals,* ed. by P. Ucko and G. W. Dimbleby. London: Duckworth, 1969.

Richards, Graham, *Human Evolution: An Introduction for the Behavioural Sciences.* London: Routledge and Kegan Paul, 1987.

Savage, R. J. G., *Mammal Evolution: An Illustrated Guide.* New York: Facts On File, 1986.

Sherratt, Andrew, ed., *The Cambridge Encyclopedia of Archaeology.* New York: Crown Publishers / Cambridge University Press, 1980.

Simons, E. L., *Primate Evolution: An Introduction to Man's Place in Nature.* London: Macmillan, 1972.

Soffer, O., and C. Gamble, eds., *The World at Eighteen Thousand BP.* London: Unwin Hyman, 1989.

Sutcliffe, Antony J., *On the Track of Ice Age Animals.* London: British Museum (Natural History), 1985.

Weisz, Paul B., *The Science of Biology.* New York: McGraw-Hill, 1963.

Willcox, A. R., *The Rock Art of Africa.* London: Croom Helm, 1984.

The World Atlas of Archaeology. London: Mitchell Beazley, 1985.

Zervos, Christian, *L'Art de l'Époque du Renne en France.* Paris: Éditions "Cahiers d'Art," 1959.

PERIODICALS

Ackerman, S., "European Prehistory Gets Even Older." *Science* (Washington, D.C.), 1989.

ApSimon, A. M., "The Last Neanderthal in France?" *Nature* (London), September 1980.

Bahn, Paul G.:
"Early Teething Troubles." *Nature* (London), 1989.
"The Unacceptable Face of the West European Upper Palaeolithic." *Antiquity* (Cambridge), 1978.

Benveniste, R. E., and G. J. Todaro, "Evolution of Type C Viral Genes: Evidence for an Asian Origin of Man." *Nature* (London), May 1976.

Bogucki, Peter, "The Antiquity of Dairying in Temperate Europe." *Expedition* (University of Pennsylvania), February 1986.

Cann, R. L., et al., "Mitochondrial DNA and Human Evolution." *Nature* (London), 1987.

COHMAP Members, "Climatic Changes of the Last 18,000 Years: Observations and Model Simulations." *Science* (Washington, D.C.), August 1988.

Diamond, J. M., "Were Neanderthals the First Humans to Bury Their Dead?" *Nature* (London), 1989.

Foley, Robert, and Robin Dunbar, "Beyond the Bones of Contention." *New Scientist* (London), October 1989.

Fricke, Hans, "Coelacanths: The Fish That Time Forgot." *National Geographic* (Washington, D.C.), June 1988.

Garrett, Wilbur E., "Where Did We Come From?" *National Geographic* (Washington, D.C.), October 1988.

Gladkih, Mikhail I., Ninelj L. Kornietz, and Olga Soffer, "Mammoth Bone Dwellings on the Russian Plain." *Scientific American* (New York), 1983.

Katz, Solomon H., and Mary M. Voight, "Bread and Beer." *Expedition* (University of Pennsylvania), February 1986.

Lewin, Roger:
"How Did Humans Evolve Big Brains?" *Science* (Washington, D.C.), 1982.
"Man the Scavenger." *Science* (London), 1984.
"Were Lucy's Feet Made for Walking?" *Science* (Washington, D.C.), 1983.

Marshack, Alexander:
"Exploring the Mind of Ice Age Man." *National Geographic* (Washington, D.C.), January 1975.
"An Ice Age Ancestor?" *National Geographic* (Washington, D.C.), October 1988.

Marshall, J. C., "The Descent of the Larynx?" *Nature* (London), 1989.

Putman, John J., "The Search for Modern Humans." *National Geographic* (Washington, D.C.), October 1988.

Renfrew, Colin, "The Origins of Indo-European Languages." *Scientific American* (New York), 1989.

Rigaud, Jean-Philippe, "Art Treasures from the Ice Age: Lascaux Cave." *National Geographic* (Washington, D.C.), October 1988.

Straus, L. G., "Age of the Modern Europeans." *Nature* (London), 1989.

Stringer, C. B., and P. Andrews, "Genetic and Fossil Evidence for the Origin of Modern Humans." *Science* (Washington, D.C.), March 1988.

Weaver, Kenneth F., "The Search for Our Ancestors." *National Geographic* (Washington, D.C.), November 1985.

Wu Rukang and Lin Shenglong, "Peking Man." *Scientific American* (New York), June 1983.

INDEX